CLUSTER ANALYSIS

A Primer Using R

CLUSTER ANALYSIS

A Primer Using R

Lior Rokach

Ben-Gurion University of the Negev, Israel

World Scientific

EW JERSEY · LONDON · SINGAPORE · BEIJING · SHANGHAI · HONG KONG · TAIPEI · CHENNAI · TOKYO

Published by

World Scientific Publishing Co. Pte. Ltd.

5 Toh Tuck Link, Singapore 596224

USA office: 27 Warren Street, Suite 401-402, Hackensack, NJ 07601

UK office: 57 Shelton Street, Covent Garden, London WC2H 9HE

Library of Congress Control Number: 2024033300

British Library Cataloguing-in-Publication Data
A catalogue record for this book is available from the British Library.

CLUSTER ANALYSIS
A Primer Using R

ISBN 978-981-12-9747-2 (hardcover)
ISBN 978-981-12-9748-9 (ebook for institutions)
ISBN 978-981-12-9749-6 (ebook for individuals)

For any available supplementary material, please visit
https://www.worldscientific.com/worldscibooks/10.1142/13968#t=suppl

Desk Editors: Soundararajan Raghuraman/Amanda Yun

Typeset by Stallion Press
Email: enquiries@stallionpress.com

Preface

In the early days of my academic career, I vividly remember a particular summer afternoon in my small, dimly lit office. My task was to make sense of a seemingly chaotic dataset of patients' electronic health records from a large hospital. The data, scattered across numerous tables, felt like an intricate labyrinth with no discernible pattern. Overwhelmed, I felt like an explorer standing at the edge of an uncharted wilderness. It was during this challenging time that I discovered the significance of cluster analysis. Cluster analysis is not just a tool for organizing data but a powerful lens through which one could see hidden patterns and relationships. By grouping patients based on their records, we were able to identify distinct segments, each with its own unique characteristics. This moment was transformative, akin to being handed a compass that could navigate through the data wilderness.

Clustering is a fundamental data analysis technique in which a set of data points are grouped into subsets, or clusters, based on their similarity. The goal of clustering is to identify natural groupings within the data, where the data points in the same cluster are more similar to each other than to data points in other clusters.

This book, "Cluster Analysis - A Primer Using R", is a comprehensive and accessible guide that takes the reader on a captivating journey through the world of data clustering, equipping them with the knowledge and skills to harness the power of this transformative analytical technique.

Cluster analysis, at its core, is the art of grouping similar data points together, revealing the inherent structure and patterns within complex datasets. In this book, readers will become familiar with the foundational principles of cluster analysis, starting with an overview of data science and data mining, followed by a deep dive into the taxonomy of machine learning tasks. This solid groundwork sets the stage for the exploration of crucial concepts, such as similarity measures, which form the backbone of the clustering process.

The heart of the book is dedicated to a thorough examination of the various clustering algorithms, spanning partitioning methods, hierarchical methods, and more advanced techniques, such as mixture density-based clustering, graph clustering, and grid-based clustering. Each method is presented with a clear and concise explanation, accompanied by illustrative examples and hands-on implementations in the R programming language, a popular and powerful tool for data analysis and visualization.

Recognizing the importance of cluster validation and evaluation, the book devotes a dedicated chapter to exploring a wide range of internal and external quality criteria, equipping readers with the necessary tools to assess the performance of clustering algorithms and determine the optimal number of clusters for their specific use cases.

For those eager to stay at the forefront of the field, Chapter 10 is dedicated to deep learning-based clustering methods, showcasing the remarkable capabilities of neural networks in uncovering hidden structures within complex, high-dimensional data.

This book, with its emphasis on fundamental principles, serves a dual purpose: it is designed as both a textbook for a one-semester elective course on cluster analysis and a comprehensive reference guide. Its well-structured content also makes it an ideal companion for machine learning courses, complementing and enhancing students' understanding of related concepts.

Whether you are a student seeking to expand your knowledge, a data analyst looking to enhance your toolbox, or a researcher exploring the frontiers of data analysis, this book will provide you with a solid foundation in cluster analysis and empower you to tackle a wide range of data-driven problems with confidence and creativity, all while leveraging the power and flexibility of the R programming language.

I would like to express my heartfelt gratitude to the entire team at World Scientific Publishing for their invaluable support and guidance throughout the creation of this book. Their expertise and dedication have been instrumental in bringing this work to fruition. Special thanks are due to Amanda Yun, Senior Editor, for her insightful feedback and unwavering encouragement, and to Soundara Rajan, Production Editor, for his meticulous attention to detail and tireless efforts in ensuring the book's polished presentation.

I owe a profound gratitude to my wife, Ronit, and my four sons, Jordan, Roy, Amit, and Michael. Their unwavering patience, boundless love, and constant support have been the pillars that sustained me through this journey.

I hope that this book will inspire you to explore the fascinating world of cluster analysis with a sense of wonder and curiosity, and that you will find the same joy in cluster analysis that I discovered all those years. Happy clustering!

Lior Rokach
Ben-Gurion University of the Negev
Beer-Sheva, Israel
August 2024

About the Author

Lior **Rokach** is a Professor of Software and Information Systems Engineering (SISE) at the Ben-Gurion University of the Negev (BGU). Professor Rokach's research group studies machine learning algorithms and their applications in Recommender Systems, Cyber Security, and Medical Informatics. Rokach has co-founded four AI companies and was awarded 22 patents for his AI and information technology inventions. From 2017 to 2020, he served as the Chair of the SISE department at BGU, and since 2023, he has served as the Vice Dean of Research for the Faculty of Engineering. Prof. Rokach is the author of over 300 peer-reviewed papers in leading journals, conference proceedings, and book chapters. Moreover, he is the author of six books, some with several editions, including *Data Mining with Decision Trees* (World Scientific, 1st edition, 2007; 2nd edition, 2015) and *Pattern Classification Using Ensemble Methods* (World Scientific, 1st edition, 2010; 2nd edition, 2019). He is the editor of the bestselling book *Recommender Systems Handbook* (Springer, 1st edition, 2011; 2nd edition, 2015; 3rd edition, 2023) and *Data Mining and Knowledge Discovery Handbook* (1st edition, Springer, 2005; 2nd edition, 2010, 3rd edition, 2023). Three of his books were translated into Chinese. He has led the development of several influential open-source software packages, most recently DeepChecks that enables researchers to thoroughly and continuously validate machine learning models. Professor Rokach currently serves as an Associate Editor for ACM Transactions on Intelligent Systems and Technology and as an Area Editor for Information Fusion. A recent study conducted by a team from Stanford University ranks Prof. Rokach among the top 0.1% of AI researchers according to their career-long impact.

Contents

Chapter 1

Introduction to Data Clustering

1.1 Overview

Cluster analysis aims to group a set of entities into clusters based on their similarity. This is a fundamental task in data analysis and data science, with applications in fields such as medicine, engineering, economics, ecology, and marketing.

The goal of this chapter is to provide the preliminary background required for reading this book. We will cover basic concepts from various interrelated fields used in the subsequent chapters. We begin by defining basic terms in data science and machine learning. Then, we provide a walk-through guide for clustering data using R.

1.2 Data Science and Data Mining

Traditionally, data collection has been recognized as a critical stage in data analysis. In the past, analysts such as statisticians or data scientists relied on their domain expertise to determine which variables should be collected. Typically, the number of variables selected was limited, and their values were often collected manually through methods such as handwritten records or oral interviews. For computer-aided analysis, the analyst would enter the collected data into a statistical computer package or electronic spreadsheet. However, due to the high cost of data collection, people were forced to make decisions based on limited information.

Since the dawn of the Information Age, it has become easier and cheaper to accumulate and store data. In fact, studies have shown that the amount of stored information doubles approximately every twenty months

[Frawley *et al.* (1992)]. However, despite the increase in machine-readable data, our ability to analyze and utilize it is not growing at the same pace.

Data Science is a field that involves the processing and analysis of data to extract useful insights. The term "Data Science" was first introduced in the 1960s, but it has gained popularity only in recent years, as technology has advanced enough to allow for efficient collection, storage, and analysis of large amounts of data. Several fields, including commerce, medicine, and research, are utilizing data-driven approaches to discover new insights and make predictions. One prominent example of a company that uses data science is Google. Google tracks user clicks to enhance the relevance of search engine results and manage its advertising campaigns more effectively.

A closely related term is data mining. Data mining refers to the process of discovering patterns or insights from large datasets by using computational algorithms and statistical techniques. With the increasing amount of data available today, data mining has become an important and necessary tool for various fields. It involves using specialized techniques and tools to automatically analyze large datasets and extract useful information. Some researchers argue that the term "Knowledge Mining" is a more accurate term as it highlights the goal of discovering valuable insights and knowledge from data, similar to the way gold is extracted from the earth through mining [Klosgen and Zytkow (2002)]. The ability to make predictions about certain phenomena is one of the ultimate goals of data science and data mining. However, prediction is a difficult task, as the famous quote attributed to Mark Twain and others goes, "It is difficult to make predictions, especially about the future". Despite this, we use prediction successfully in various applications. For example, YouTube, a popular website owned by Google, analyzes users' viewing habits to predict which other videos they may be interested in. Based on this prediction, YouTube provides personalized recommendations, which are often effective. To roughly estimate the service's efficiency, one could ask oneself how often watching a video on YouTube leads to watching similar videos recommended by the system. Similarly, online social networks (OSNs) such as Facebook and LinkedIn automatically suggest friends and acquaintances with whom users can connect.

A common question that arises is how data science differs from statistics and what sets data scientists apart from statisticians. Data science takes a holistic approach as it encompasses the entire data analysis process, from data sensing and collection to storage, processing, feature extraction, data mining, and knowledge discovery. This interdisciplinary field incorporates theories and techniques from multiple domains, including statistics, mathematics, computer science, and especially sub-fields like Artificial Intelligence and information technology.

1.3 The Four-Layers Model

Arranging the data mining domain into four layers is useful. Figure 1.1 presents this model. The first layer represents the target application. Data mining can benefit many applications, such as:

(1) Credit Scoring — The aim of this application is to evaluate the creditworthiness of a potential consumer. Banks and other companies use credit scores to estimate the risk posed by doing a business transaction (such as lending money) with this consumer.

(2) Fraud Detection — Oxford English Dictionary defines fraud as "An act or instance of deception, an artifice by which the right or interest of another is injured, a dishonest trick or stratagem". Fraud detection aims to identify fraud as quickly as possible once it has been perpetrated.

(3) Churn Detection — This application helps sellers identify customers with a higher probability of leaving and potentially moving to a competitor. By identifying these customers in advance, the company can act to prevent churning (for example, by offering a better deal to the consumer).

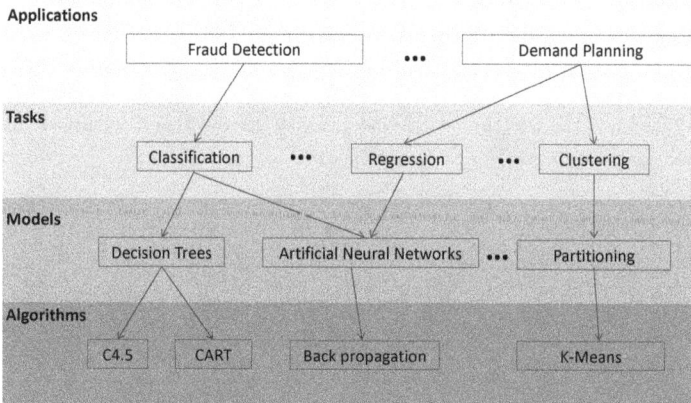

Fig. 1.1: The four layers of data mining.

Each application is built by accomplishing one or more machine-learning tasks. The second layer in our four-layer model is dedicated to the machine learning tasks, such as: Classification, Clustering, Anomaly Detection, Regression etc. Each machine-learning task can be accomplished by various machine-learning models, as indicated in the third layer. For example, the classification task can be accomplished by the following two models: Decision Trees or Artificial Neural Networks. In turn, each model can be induced from the training data using various learning algorithms. For example, a decision tree can be built using either the C4.5 algorithm or the CART algorithm, which will be described in the following chapters.

1.4   Taxonomy of Machine Learning Tasks

In the machine learning community, it is common to distinguish between two main types of machine learning tasks: Supervised learning and unsupervised learning. Supervised learning methods aim to discover patterns from labeled training data in an attempt to model the relationship between input attributes (sometimes called independent variables) and a target attribute (sometimes referred to as a dependent variable). Approaches to supervised learning include: classification and regression. Unsupervised learning methods, on the other hand, refer to methods that aim to discover data-driven patterns in unlabeled data, i.e., without a pre-specified dependent variable. Figure 1.2 illustrates this taxonomy.

Clustering analysis is considered to be an unsupervised task because we do not know in advance which entities belong to the same cluster nor how many clusters exist in the data. In essence, the difference between clustering and classification lies in the custom in which knowledge is extracted from data: the goal of clustering is descriptive, whereas the goal of classification is predictive. Another distinction between clustering and classification stems from their different goals. This difference, as indicated by [Veyssieres and Plant (1998)], involves the way that the groups resulting from each sort of analysis are assessed: Since the goal of clustering is to discover a new set of categories, the new groups are of interest in themselves, and their assessment is intrinsic. In classification tasks, however, an important part of the assessment is extrinsic, since the groups must reflect some reference set of classes.

"Understanding our world requires conceptualizing the similarities and differences between the entities that compose it" [Tryon and Bailey (1970)].

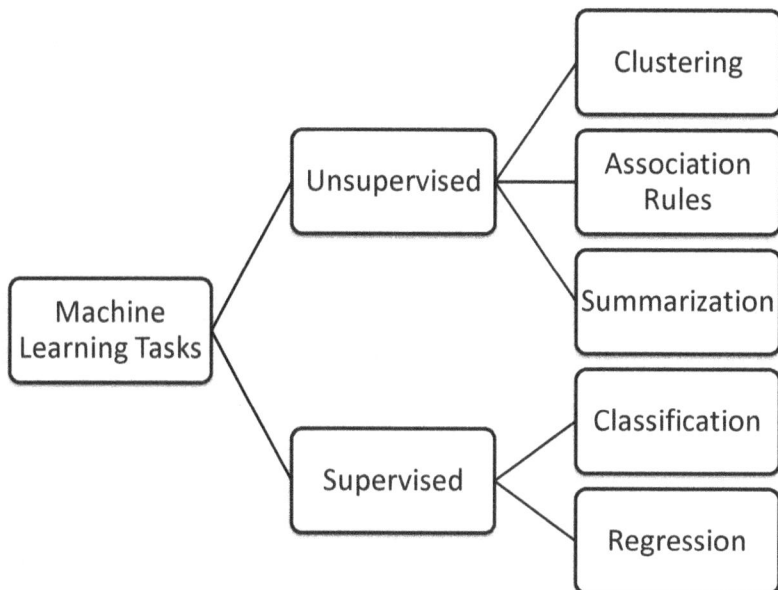

Fig. 1.2: Taxonomy of machine learning methods.

Clustering does exactly so. It groups data instances into subsets, ensuring that similar instances are grouped together while different instances are placed in separate groups. This process organizes the instances into an efficient representation that characterizes the sampled population.

1.4.1 *Data representation*

In a typical clustering task, the main input of the algorithm is a dataset that describes the objects to be clustered. The dataset can be described in a variety of languages. Most frequently, it is described as a $m \times n$ matrix where the rows correspond to the objects and the columns to their characteristics. Each object (also known as sample, record, or instance) is described by a vector of n attribute values. Attributes (sometimes called measurement, field, variable or feature) are typically one of two types: nominal (values are members of an unordered set), or numeric (values are real numbers).

Table 1.1: Illustration of a dataset having five attributes.

a_1	a_2	a_3	a_4	a_5
Yes	17	4	7	0
No	81	1	9	1
Yes	17	4	9	0
No	671	5	2	0
Yes	1	123	2	0
Yes	1	5	22	1
No	6	62	1	1
No	6	58	54	0
No	16	6	3	0

Let A denotes the set of input attributes containing n attributes: $A = \{a_1, \ldots, a_i, \ldots, a_n\}$. When the attribute a_i is nominal, it is useful to denote by $dom(a_i) = \{v_{i,1}, v_{i,2}, \ldots, v_{i,|dom(a_i)|}\}$ its domain values, where $|dom(a_i)|$ stands for its finite cardinality. Numeric attributes have infinite cardinalities. The instance space (the set of all possible examples) is defined as a Cartesian product of all the input attributes domains: $X = dom(a_1) \times dom(a_2) \times \cdots \times dom(a_n)$.

In order to refer to the ith object in the dataset, we use the notation $\mathbf{x_i}$ that represents a n-length vector. When the value of its jth attribute is denoted by $x_{i,j}$ Take, for example, the dataset presented in Table 1.1. It contains nine objects, and each one of them is described by five attributes. The first attribute (a_1) is nominal, and the rest are numeric. The first object is represented by $\mathbf{x_1} = (Yes\ 17\ 4\ 7\ 0)$.

1.5   What is Clustering?

Three common definitions for clustering are given in the literature:

(1) "*Cluster is a set of entities which are alike, and entities from different clusters are not alike*". [Jain et al. (1988)]
(2) "*An aggregate of points in the test space such that the distance between any two points in the cluster is less than the distance between any point in the cluster and any point not in it*". [Gengerelli (1963)]
(3) "*Clusters may be described as connected regions of a multi-dimensional space containing a relatively high density of points, separated from other such regions by a region containing a relatively low density of points*". [Jain et al. (1988)]

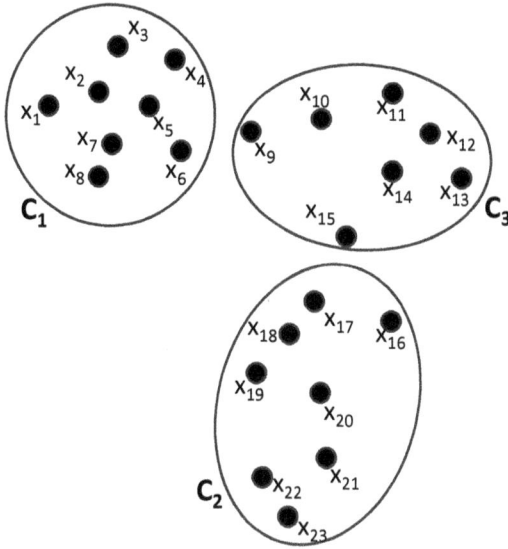

Fig. 1.3: Illustration of a hard partitioning.

Figure 1.3 illustrates a *hard* clustering that consists of three clusters $\mathbf{C} = (C_1, C_2, C_3)$. Each data instance is associated with exactly one cluster. Figure 1.4 illustrates a *soft* partitioning with three clusters (C_1, C_2 and C_3). Note that data instances $\mathbf{x_9}$ and $\mathbf{x_{10}}$ are associated with two clusters (C_1 and C_3). Moreover data instance $\mathbf{x_{15}}$ is associated with clusters C_2 and C_3.

Figure 1.5 illustrates a *hierarchical* clustering. Cluster C_0 consists of all data instances. It is divided into two clusters C_1, which consists of instances $\mathbf{x_1}$ to $\mathbf{x_{15}}$, and C_2 which includes all remaining instances. Cluster C_1 is further divided into clusters C_5 and C_6 while cluster C_2 is further divided into clusters C_3 and C_4. Figure 1.6 presents the hierarchical structure of the clusters presented in Figure 1.5.

Next, we provide a formal definition of the abovementioned clustering structure. Given a dataset of m entities $\mathbf{X} = \{\mathbf{x_1}, \mathbf{x_2}, \ldots, \mathbf{x_m}\}$ where each entity is described by a n-sized vector $\mathbf{x_i} = (x_{i1}, x_{i2}, \ldots, x_{in})$, the following clustering structures can be defined:

(1) Hard partitioning is defined as a set of clusters $\mathbf{X} = \{C_1, C_2, \ldots, C_k\}$ such that $\mathbf{X} = \bigcup_{i=1}^{k} C_i$ and $C_i \cap C_j = \emptyset; \; \forall i \neq j$ where k is the number of clusters. This means that every entity belongs to exactly one cluster.

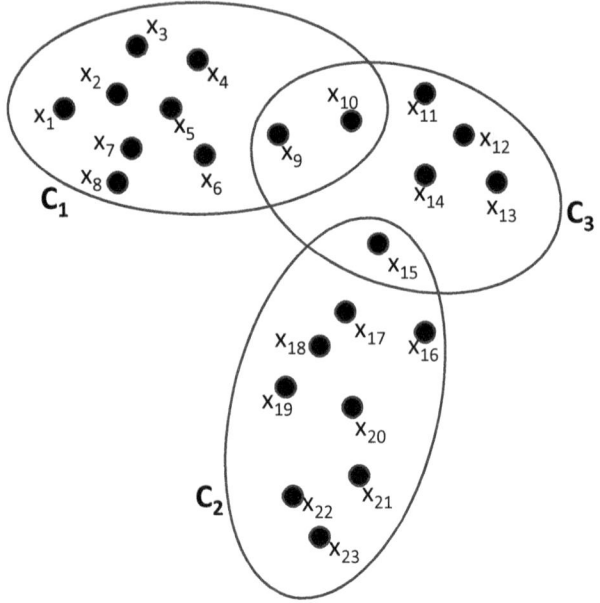

Fig. 1.4: Illustration of a soft partitioning.

(2) Soft (fuzzy) partitioning is defined as a membership matrix $U \in \mathbb{R}^{mk}$ such that $u_{ij} \in [0,1]$; ; $\forall i, j$ and $\sum_{j=1}^{k} u_{ij} = 1$; ; $\forall i$ where u_{ij} represents the membership value of entity i regarding cluster j. Note that hard partition can be regarded as a special case of fuzzy partition, in which every entity is associated with a membership value of 1 to exactly one cluster and 0 to all other clusters.

(3) Hierarchical clustering is defined as a set of clusters $\mathbf{X} = \{C_0, C_1, C_2, \ldots, C_k\}$ such that $C_0 \triangleq \mathbf{X}$ and $\forall i \neq j$ one of the following conditions is true: $C_i \subset C_j$ or $C_j \subset C_i$ or $C_i \bigcap C_j = \emptyset$. In addition for any cluster C_i for which $\exists C_j \subset C_i$, there exists a set of disjoint clusters $\{C_j\}_{j=1}^{r}$ such that $\bigcup_{j=1}^{r} C_j = C_i$.

In the last two sections, we have seen that the main terms in cluster analysis have several synonyms. Table 1.2 specifies the common synonyms for these terms.

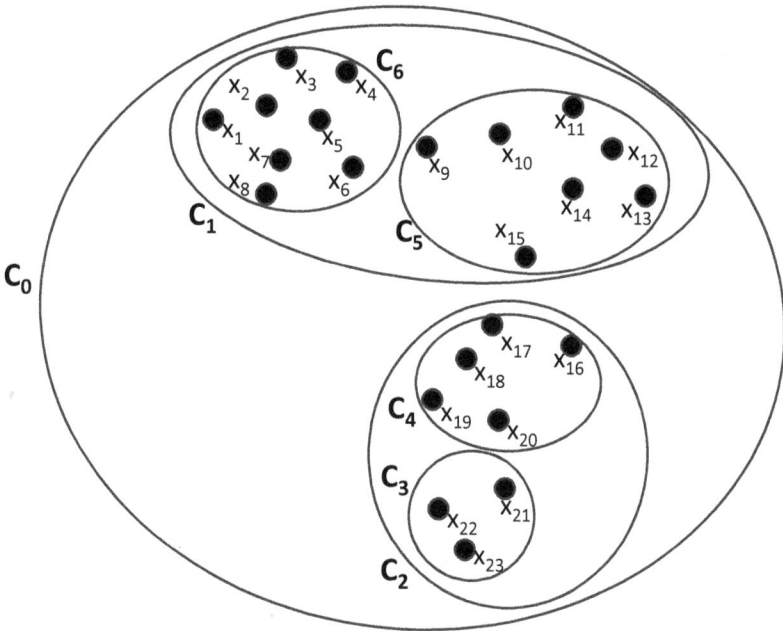

Fig. 1.5: Illustration of a hierarchical clustering.

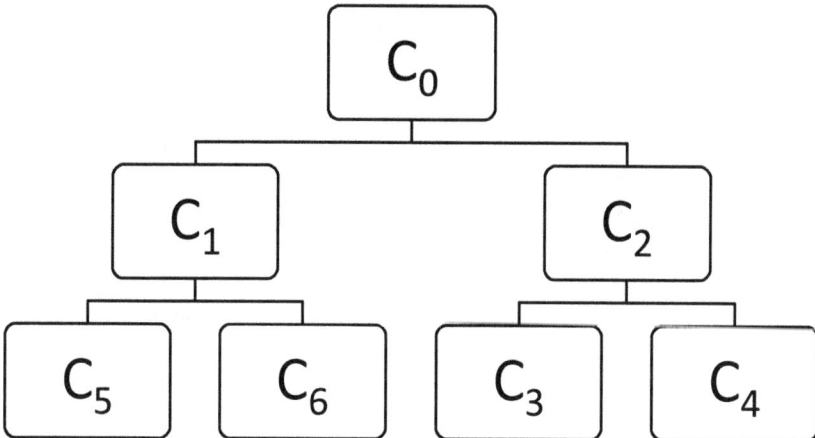

Fig. 1.6: The hierarchical clustering structure.

Table 1.2: Common synonyms for main terms in cluster analysis.

Term	*Synonym*
Cluster	Groups, subsets, categories
Entities	Objects, patterns, instances, observations, units, samples, feature vectors, tuples
Attribute	Feature, dimension, variable

1.6   Taxonomy of Clustering Methods

In this section, we further explore the clustering node in the machine learning taxonomy presented in Figure 1.2. Recall we opened this chapter by
saying that cluster analysis aims to group a set of entities into clusters in
such a manner that entities that have a high degree of similarity are grouped
together. However, this is still a vague definition mainly because similarity
is considered to be a subjective issue. For example, if two unrelated people
are asked to cluster into two groups the characters of the well-known ani-
mated sitcom, The Simpsons; we should expect to get different and equally
correct clusterings, e.g., clustering by gender (males vs. females), clustering by age (children vs. adults) etc. In fact, there is no precise definition
for the notion of a "cluster" nor is there a universally agreed definition.
Therefore, many researchers agree with the statement that "clustering is in
the eye of the beholder" [Estivill-Castro (2002)].

This vague definition of cluster analysis has led to the development
of many clustering methods. It is useful to organize these methods in a
taxonomy based on shared characteristics.

In this book, we propose the taxonomy presented in Figure 1.7. While
there is no canonical taxonomy for clustering methods, many researchers
agree upon two basic types of clustering methods: partitioning methods
and hierarchical methods. Partitioning methods discover clusters mainly by
identifying subspaces densely populated with entities. On the other hand,
hierarchical methods gradually assemble entities into clusters according to
some dissimilarity measure between any two entities.

Each basic type can be further divided into sub-types as illustrated in
Figure 1.7. Specifically, hierarchical methods can be sub-divided as follows:

- Agglomerative hierarchical clustering — It begins with each object as
  its own individual cluster. Clusters are then progressively merged until
  the desired cluster structure is achieved.

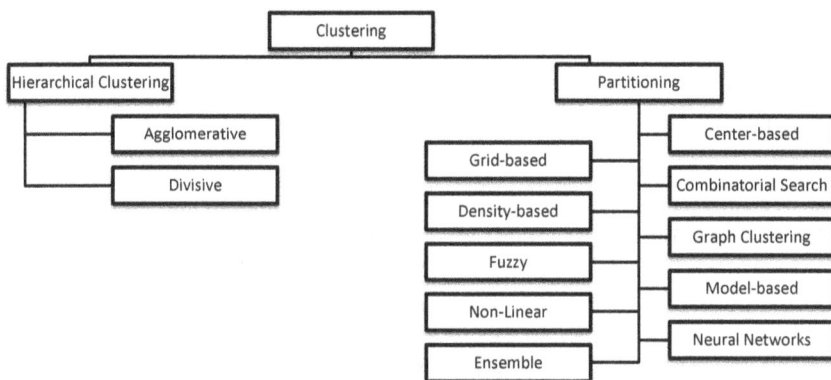

Fig. 1.7: Taxonomy of clustering methods.

- **Divisive hierarchical clustering** — All objects initially belong to one cluster. Then, the cluster is divided into sub-clusters, which are successively divided into sub-clusters of their own. This process continues until the desired cluster structure is obtained.

Partitioning methods can be sub-divided as follows:

- **Center-based** — The main concept is to identify a clustering structure that minimizes an error criterion measuring the "distance" of each instance from its representative value. The sum of the squared error function is a prevalent criterion used to find the optimal partition. These methods, effective for isolated and compact clusters, are both intuitive and commonly employed.
- **Combinatorial search-based methods** — A clustering task can be defined as an optimization problem that aim to organize the entities into subsets that optimize some criterion function. The idea of combinatorial search methods is to approximate global optimal clustering structure by efficiently exploring the large solution space instead of traversing through all possible partitions.
- **Graph-theoretic clustering** — Graph theoretic methods are methods that produce clusters via graphs. The edges of the graph connect the entities, which are represented as nodes. The edges reflect the proximities between each pair of entities. Thus, graph clustering aims to partition a graph into several densely connected subgraphs such that nodes within the same cluster have more connections than those in different clusters.

- Model-based methods operate on the assumption that points within each cluster are derived from a specific probability distribution. The overall data distribution is considered a mixture of several such distributions. These methods aim to identify the clusters and determine their distribution parameters. They are capable of detecting clusters of arbitrary shapes, which do not need to be convex.

- Neural networks-based method — This algorithm represents each cluster with a neuron or prototype. Neurons also represent the input data, and these are connected to the prototype neurons. Each connection has a weight that is adaptively learned during the training process.

- Grid-based — These methods first use a grid mesh to partition the entire input space into cells, then calculate statistic features for each cell and, finally, perform the clustering on the grid instead of the data instances directly. These methods are characterized by a low computational cost and, therefore, are suitable for very large data sets.

- Density-based — These methods aim to find arbitrarily shaped clusters by looking for high-density regions separated by low-density regions. Since these methods automatically identify high-density regions, the user is not required to predefine the number of clusters.

- Fuzzy clustering — In hard clustering, every entity belongs to exactly one cluster. In Fuzzy clustering, entities may be associated with more than one cluster according to some affinity level, which indicates the strength of the association between that entity and a particular cluster. A fuzzy clustering can be converted to a hard clustering by assigning each entity to the cluster with the largest measure of membership.

- Nonlinear partitioning methods are able to produce nonlinear separating hyper-surfaces between clusters. This class of algorithms consists of two main approaches: Spectral methods and kernel methods. Spectral clustering methods aim to reduce the original dimensionality of the data before performing the actual clustering. Here, one utilizes the top eigenvalues of the similarity matrix among the entities. Kernel clustering methods aim to map the original data into a high-dimensional space. Then, they search for linear partitioning in this new space, which corresponds to nonlinear partitioning in the original space.

- Cluster ensemble aims to improve the quality of clustering by first generating a set of clusterings using the same data (e.g., by using different clustering algorithms) and then combining them into a new final clustering, which is the integration of all partitions generated.

1.7 Data Clustering Using R

R is a free software programming language widely utilized by data scientists to develop data mining algorithms. The extensive and diverse range of contributed packages available in the Comprehensive R Archive Network (CRAN) allows most data science tasks to be efficiently completed with concise script code in R.

The strength of R comes from the various functions which are provided by different freely available packages. Specifically, several packages offer the implementation of various clustering methods. In this section, we will focus on the clustering methods that come with the *stats* package of R. Stats is a basic package that is automatically loaded at the start of an R session. It provides a broad range of statistical functionality, including the implementation of the two most popular clustering algorithms: *kmeans* and *hclust*. The usage of these two methods is described below.

Code 1.1 illustrates the usage of *kmeans* function. It is mandatory to provide the data to be clustered (x) and either the number of clusters (k) or the coordinates of the initial set of cluster centers. The k-means method partitions the data instances given by x into k groups such that the sum squared distances of the data instances to its closest cluster center is minimized. The *kmeans* method includes other controlling parameters that are optional, such as iter.max, which specifies the maximum number of iterations allowed, and the algorithm, which may be one of the following: "Lloyd" (the default algorithm), "Hartigan-Wong" or "MacQueen".

In the following example, we are using the *USJudgeRatings* dataset. This dataset is available as part of R (the dataset package), which includes the Lawyers' Ratings of State Judges in the US Superior Court. There are 43 rows and 12 columns. Each row refers to a different judge. Each column refers to a different numeric criterion, such as prompt decisions and preparation for trial. For the sake of 2D visualization, we will use only the first two criteria: *CONT* (the number of contacts of a lawyer with the judge) and *INTG* (the judicial integrity). Line 2 of Code 1.1 projects the original dataset into a 2D dataset using only the first two columns. In line 3, the k-means clustering is performed using $k = 2$. In line 4, the 3D dataset is plotted such that the point's symbol is determined based on its clusters assignment. Finally, in line 5, the centroids of the clusters are displayed. The resulting graph is presented in Figure 1.8.

Listing 1.1: Using kmeans to cluster USJudgeRatings dataset

```
1 x=USJudgeRatings[,1:2]
2 cl<-kmeans(x,2)
3 plot(x, pch = cl$cluster)
4 points(cl$centers, pch = 1:2, cex = 2)
```

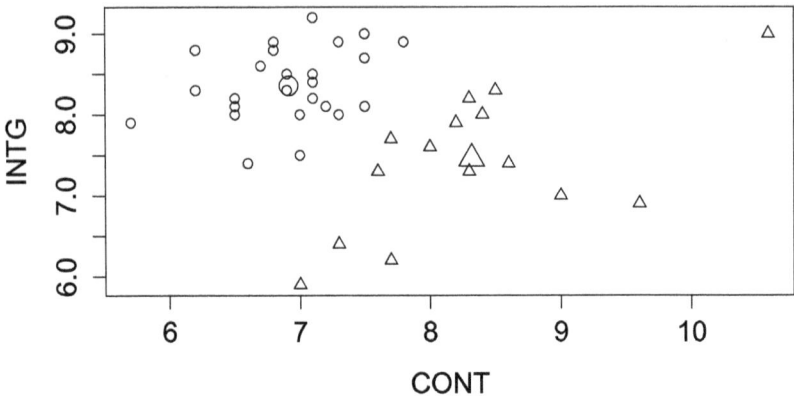

Fig. 1.8: Output of Code 1.1.

In the next code example, we are using the basic hierarchical clustering algorithm to cluster the same emphUSJudgeRatings dataset. As in the previous example, line 2 of Code 1.2 projects the original dataset into a 2D dataset using only the first two columns. In line 3, the distance matrix between any pair of judges is computed using the Euclidean distance function (more on this issue in the next chapter). In line 4, the hierarchical clustering is performed using the average agglomeration method. The algorithm begins with assigning each judge to its own cluster. Then, at each iteration, the algorithm joins the most similar clusters to a single cluster. The iterations continue till we are left with a single cluster, which consists of all judges in the dataset, In line 5, the cluster tree, or dendrogram, is plotted. The resulting dendrogram is presented in Figure 1.9. It illustrates the hierarchical arrangement of the clusters produced by the clustering. The leaves of the dendrogram (the lowest nodes in the graph) represent the

Listing 1.2: Using hierarchical clustering to cluster USJudgeRatings dataset

```
1 x=USJudgeRatings[,1:2]
2 DistX=dist(x)
3 cl<-hclust(DistX,"ave")
4 plot(cl)
```

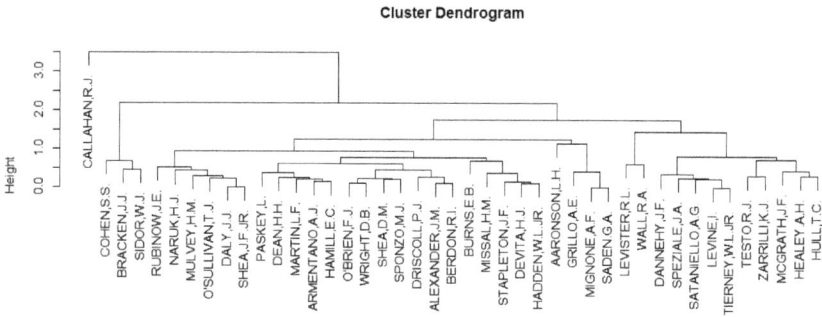

Fig. 1.9: Output of Code 1.2.

individual judges. The remaining nodes represent the clusters to which the judges are assigned. The root of the dendrogram (the node located at the top of the graph) represents the universal cluster that consists of all judges in the dataset. The height of each node in the graph refers to the distance between its two children.

Chapter 2

Similarity Measures

2.1 Overview

Since clustering involves grouping similar instances/objects, a measure is needed to determine their similarity or dissimilarity. There are two primary types of measures used to estimate this relation: Distance Measures and Similarity Measures. This chapter presents commonly used similarity and dissimilarity measures and how they can be used in R.

2.2 Preliminaries

2.2.1 Data types

Before we can define the distance measure among objects, we should first get familiar with the existing different attribute types:

- Numerical attribute is an attribute that is used to represent observations' values that can be quantitatively measured. Sometimes, it is helpful to further divide the numeric attributes into two sub-types: real (such as a person's weight) and count (such as the number of items a customer purchased). A count attribute can take only non-negative integer values $\{0, 1, 2, 3, \ldots\}$. An easy way to decide if an attribute is numerical is to examine if it is possible to calculate an average or sum of all values of this attribute. If possible, then the attribute is a numerical variable.

  It is also useful to differentiate between the following two sub-types of numerical attributes: interval scale and ratio scale. An interval scale attribute is a measurement where the difference between two values is meaningful but not the ratio between them. Moreover, interval scale does not have a meaningful zero value. On the other hand, in ratio scale attributes, the ratio of two values is meaningful. A ratio scale possesses a meaningful absolute zero value that represents the "Real Origin".

For example, an attribute that represents a temperature expressed in Celsius is an interval scale attribute. First, a temperature of $0°C$ does not mean "no heat". Second, The interval difference between $2°C$ and $4°C$ is the same as between $4°C$ and $6°C$. However, the temperature $2°C$ is not twice the air heat energy as $1°C$. On the other hand, an attribute that represents a temperature expressed in Kelvin is a ratio scale attribute because it has an absolute non-arbitrary zero value at which substances pa ossess no thermal energy.

- Categorical attribute (also known as nominal attribute) is a qualitative attribute that can take on one of a limited, and usually fixed, number of possible values (categories), and there is no clear ordering to the categories. For example, tracking the manufacturer of the customer's smartphone might be useful in a customer analytic scenario. In this case, the feature *smartphone manufacturer* is a categorical attribute having several categories (Samsung, HTC, Huawei) with no intrinsic ordering among the categories.

- Binary attribute is a categorical attribute that can take only two possible values. Even though binary attributes are frequently coded numerically as 0 and 1, they cannot be compared numerically. Because the values 0 and 1 stand for qualitatively different values.

- Ordinal attribute is a qualitative attribute in which the order of the values is no longer arbitrary, unlike a categorial attribute. Instead, the values' order does matter, but not the difference between the values. For example, the membership level of a credit card holder can be assigned to the following ordered levels: Ivory, Silver, Gold, Platinum. Obviously the Platinum level is considered to be better than the Gold level and the Gold level is considered to be better than the Silver level and so forth. However, the differences between the levels cannot by quantitatively measured.

2.2.2   *Distance measures*

Many clustering methods use distance measures to assess the dissimilarity between pairs of objects. Given a dataset of m entities $\mathbf{X} = \{\mathbf{x_1}, \mathbf{x_2}, \ldots, \mathbf{x_m}\}$, it is useful to denote the distance between two objects $\mathbf{x_i}$

and $\mathbf{x_j}$ as: $d(\mathbf{x_i}, \mathbf{x_j})$. Recall that $\mathbf{x_i}$ and $\mathbf{x_j}$ are both n-length vectors. Thus, the distance is defined as some function of the attributes' values of the two objects, i.e.:

$$d(\mathbf{x_i}, \mathbf{x_j}) = d(x_{i,1}, x_{i,2}, \ldots, x_{i,n}, x_{j,1}, x_{j,2}, \ldots, x_{j,n}) \tag{2.1}$$

A valid distance measure should be symmetric, i.e. $d(\mathbf{x_i}, \mathbf{x_j}) = d(\mathbf{x_j}, \mathbf{x_i})$ and obtains its minimum value of 0 in case of identical vectors, i.e. $d(\mathbf{x_i}, \mathbf{x_i}) = 0$. The distance measure is called a metric distance measure, if it satisfies the following properties:

(1) Triangle inequality: $d(\mathbf{x_i}, \mathbf{x_k}) \leq d(\mathbf{x_i}, \mathbf{x_j}) + d(\mathbf{x_j}, \mathbf{x_k}) \quad \forall \mathbf{x_i}, \mathbf{x_j}, \mathbf{x_k} \in \mathbf{X}$.
(2) Reflexivity: $d(\mathbf{x_i}, \mathbf{x_j}) = 0 \iff \mathbf{x_i} = \mathbf{x_j} \quad \forall \mathbf{x_i}, \mathbf{x_j} \in \mathbf{X}$.
(3) Nonnegativity: $d(\mathbf{x_i}, \mathbf{x_j}) \geq 0 \quad \forall \mathbf{x_i}, \mathbf{x_j} \in \mathbf{X}$.
(4) Commutativity: $d(\mathbf{x_i}, \mathbf{x_j}) = d(\mathbf{x_j}, \mathbf{x_i}) \quad \forall \mathbf{x_i}, \mathbf{x_j} \in \mathbf{X}$.

A distance matrix or dissimilarity matrix is a square matrix that consists of the pairwise distance measures of all objects in the dataset. Distance matrix is not necessarily hollow. That being said, in most cases it is indeed hollow with all diagonal elements equal to zero. The distance matrix is symmetric when the distance measure complies with the commutativity property.

Given a dataset of m entities $\mathbf{X} = \{\mathbf{x_1}, \mathbf{x_2}, \ldots, \mathbf{x_m}\}$. Each entity $\mathbf{x}_i$ is a n-dimensional vector. The distance matrix M_{dist} for $\mathbf{X}$ is defined as:

$$M_{dist} = \begin{pmatrix} d(\mathbf{x_1}, \mathbf{x_1}) & d(\mathbf{x_1}, \mathbf{x_2}) & \cdots & d(\mathbf{x_1}, \mathbf{x_m}) \\ d(\mathbf{x_2}, \mathbf{x_1}) & d(\mathbf{x_2}, \mathbf{x_2}) & \cdots & d(\mathbf{x_2}, \mathbf{x_m}) \\ \vdots & \vdots & \ddots & \vdots \\ d(\mathbf{x_m}, \mathbf{x_1}) & d(\mathbf{x_m}, \mathbf{x_2}) & \cdots & d(\mathbf{x_m}, \mathbf{x_m}) \end{pmatrix} \tag{2.2}$$

2.3 Euclidean Distance

The Euclidean distance is probably the most popular distance measure in the clustering literature. It is simply defined as the straight line distance between two points in the Euclidean space. The two objects must have only numeric attributes. Euclidean distance is also known by the name l_2 norm.

Given two n-dimensional instances: $\mathbf{x_i} = (x_{i1}, x_{i2}, \ldots, x_{in})$ and $\mathbf{x_j} = (x_{j1}, x_{j2}, \ldots, x_{jn})$, the Euclidean distance between the two instances is defined as:

$$d(\mathbf{x_i}, \mathbf{x_j}) = \sqrt{(x_{i1} - x_{j1})^2 + (x_{i2} - x_{j2})^2 + \cdots + (x_{in} - x_{jn})^2} \qquad (2.3)$$

A special attention should be paid to the fact that in the plain Euclidean distance the values on each dimension are not standardized. Thus, if the values scale of one attribute is much larger than the other variables, then the distance measure will be greatly biased toward the attribute that have the largest scale. The weighted Euclidean distance allows to incorporate nonnegative weights to all attributes, and by that control the importance of each attribute:

$$d(\mathbf{x_i}, \mathbf{x_j}) = \sqrt{w_1 \times (x_{i1} - x_{j1})^2 + \cdots + w_n \times (x_{in} - x_{jn})^2} \qquad (2.4)$$

In terms of the last equation, the plain Euclidean distance simply assumes that all attributes are weighted equally with $w_i = 1$.

2.4   Minkowski: Distance Measures for Numeric Attributes

Given two n-dimensional instances: $\mathbf{x_i} = (x_{i1}, x_{i2}, \ldots, x_{in})$ and $\mathbf{x_j} = (x_{j1}, x_{j2}, \ldots, x_{jn})$, the distance between the two data instances can be calculated using the Minkowski metric (named after the mathematician Hermann Minkowski) [Han *et al.* (2011)]:

$$d(\mathbf{x_i}, \mathbf{x_j}) = (|x_{i1} - x_{j1}|^g + |x_{i2} - x_{j2}|^g + \cdots + |x_{in} - x_{jn}|^g)^{1/g} \qquad (2.5)$$

where g is the order of the Minkowski distance. For $g \geq 1$, the Minkowski distance is a metric because it complies with the triangle inequality. This is not true for $g < 1$.

The abovementioned Euclidean distance is a special case of Minkowski distance when $g = 2$. Given $g = 1$, the sum of absolute paraxial distances is obtained. This is known by the name Manhattan distance or by the name l_1 norm. With $g = \infty$ one gets the greatest of the paraxial distances which is usually referred to as Chebychev distance.

Recall that each attribute's measurement unit can affect the clustering analysis. To ensure consistency regardless of measurement units, it is important to standardize the data. Standardizing measurements aims to assign equal weight to all variables. However, if each variable is assigned with a weight according to its importance, then the weighted Minkowski distance is defined as:

$$d(\mathbf{x_i}, \mathbf{x_j}) = (w_1\,|x_{i1} - x_{j1}|^g + w_2\,|x_{i2} - x_{j2}|^g + \cdots + w_n\,|x_{in} - x_{jn}|^g)^{1/g}$$
(2.6)

where $w_i \in [0, \infty)$.

2.5 Distance Measures for Binary Attributes

The distance measure described in the last section may be easily computed for numerical variables. In the case of instances described by categorical, binary, ordinal or mixed type attributes, the distance measure should be adjusted accordingly.

If objects are represented by n-length binary vectors, then the distance between two such objects may be calculated based on a contingency table, as shown in Table 2.1.

Table 2.1: Contingency table for binary attributes.

		Instance j		
		1	0	sum
Instance i	1	q	r	q+r
	0	s	t	s+t
	Sum	q+s	r+t	n

where q is the number of variables that equal 1 for both objects, t is the number of variables that equal 0 for both objects, and s and r are the number of attributes that are non equal for both attributes. The total number of attributes is n.

A binary attribute is symmetric, if both of its states can be assumed to have equal relevance. For example an attribute the represents the gender of the customer is considered to be symmetric, because both values "male" and "female" are equally important. In that case, using the simple matching coefficient can assess dissimilarity between two objects:

$$d(\mathbf{x_i}, \mathbf{x_j}) = \frac{r + s}{q + r + s + t}$$
(2.7)

Equation (2.7) is referred in the literature as *afinity index*. It became the most common similarity measure for binary attribute due to its simplicity and it intuitive nature. Still in certain cases other measures are used, such as: Rogers-Tanimoto distance in which the disagreements term $r + s$ carry twice the weight of the agreements term $(q + t)$:

$$d(\mathbf{x_i}, \mathbf{x_j}) = \frac{2(r + s)}{q + t + 2(r + s)} \tag{2.8}$$

Similarly Sokal-Sneath distance puts the double weight on the agreement term, i.e.:

$$d(\mathbf{x_i}, \mathbf{x_j}) = \frac{r + s}{2(q + t) + r + s} \tag{2.9}$$

A binary attribute is asymmetric, if its states are not equally important (usually the positive outcome considered more important). For example, an attribute can be used to represents if a customer has previously read s a certain book (True – for reading the book, False – for not reading the book). In such a case, two readers have a similar taste if they both read a certain book (i.e. both have the value True). On the other hand, we cannot derive the same conclusion if both have not read a certain book (i.e. both have the value False). Thus, in this case the value "True" is more important for determining the similarity of the two readers. In this case the similarity formula presented in Equation (2.7) should be adjusted such that denominator ignores the unimportant negative matches (t). This is called the Jaccard coefficient:

$$d(\mathbf{x_i}, \mathbf{x_j}) = \frac{r + s}{q + r + s} \tag{2.10}$$

2.6   Distance Measures for Categorical Attributes

When the attributes are *Categorical*, two main approaches may be used:

(1) Simple matching:

$$d(\mathbf{x_i}, \mathbf{x_j}) = \frac{n - t}{n}$$

where n is the total number of attributes and t is the number of matches that is, the number of attributes for which instances i and j happen to have the same category value.

(2) Creating a binary attribute for each state of each nominal attribute, and computing their dissimilarity as described in Equation (2.10).

2.6.1 *Distance metrics for ordinal attributes*

When the attributes are *ordinal* the order of the values are meaningful. It would be unwise to address an ordinal attribute as if it was a categorical attribute, because the closer two values are, the shorter the resulting distance should become. In such cases, the attributes can be treated as numeric ones after mapping their range onto $[0, 1]$. Such mapping may be carried out as follows:

$$z_{i,k} = \frac{r_{i,k} - 1}{M_k - 1} \tag{2.11}$$

where $z_{i,k}$ is the standardized value of attribute a_k of object i. The function $r_{i,k}$ corresponds to the rank of the value $x_{i,k}$ in the ordered list of values (assuming the first value gets the value of 1) and M_k is the length of the list. Recall the example in which a customer's attribute holds the membership level of a credit card holder using the following ordered levels: Ivory, Silver, Gold, and Platinum. A customer with a value of "Gold" will be converted to a numeric value as follows:

$$\frac{3 - 1}{4 - 1} = 0.75 \tag{2.12}$$

Note that the rank of "Gold" in the ordered list {Ivory, Silver, Gold, Platinum} is three and the length of the list is 4.

Once the ordinal attributes have been converted to numerical attributes, we can use all existing distance metrics, such as Euclidean distance, to calculate the distance between two instances.

2.7 Distance Metrics for Mixed-Type Attributes

In the case that the instances are characterized by attributes of *mixed-type*, one may calculate the distance by combining the methods mentioned above. For instance, when calculating the distance between instances i and j using a metric such as the Euclidean distance, one may calculate the difference between nominal and binary attributes as 0 or 1 ("match" or "mismatch", respectively), and the difference between numeric attributes as the difference between their normalized values. The square of each such difference will be added to the total distance. Such calculation is employed in many clustering algorithms presented later.

The *Gower* distance $d(\mathbf{x_i}, \mathbf{x_j})$ between two instances, containing n attributes of mixed types, are defined as:

$$d(\mathbf{x_i}, \mathbf{x_j}) = \frac{\sum_{k=1}^{n} \delta_{ij}^{(k)} d_{ij}^{(k)}}{\sum_{k=1}^{n} \delta_{ij}^{(k)}} \tag{2.13}$$

where the indicator $\delta_{ij}^{(k)}$ gets the value 0, if one of the values is missing, and it is put equal to 1 otherwise. In addition, $\delta_{ij}^{(k)}$ is set to 0 when attribute a_k is an asymmetric binary attribute and instances i and j both have a value of 0.

The contribution of attribute k to the distance between the two objects $d^{(k)}(\mathbf{x_i}, \mathbf{x_j})$ is computed according to its type:

- If the attribute k is binary or categorical, $d^{(k)}(\mathbf{x_i}, \mathbf{x_j}) = 0$ if $x_{ik} = x_{jk}$, otherwise $d^{(k)}(\mathbf{x_i}, \mathbf{x_j}) = 1$.
- If the attribute is an interval scale numerical attribute, then $d_{ij}^{(k)} = \frac{|x_{ik} - x_{jk}|}{\max_h x_{hk} - \min_h x_{hk}}$, where h runs over all instances with non-missing values for attribute k. Thus, $\max_h x_{hk} - \min_h x_{hk}$ represents the range of attribute a_k.
- if the attribute is a ratio scale numerical attribute, then the values are first converted to interval scale numerical values by carrying out logarithmic transformation and then addressed as an interval scale numerical attribute.
- If the attribute is ordinal, the standardized values of the attribute are computed first, and then, $z_{i,k}$ is treated as an interval scale numerical attribute.

Note that because $d_{ij}^{(k)}$ is restricted to the range $[0, 1]$ for each k, the resulting distance $d(\mathbf{x_i}, \mathbf{x_j})$ also restricted to the range $[0, 1]$.

The advantage of the Gower distance is that it is easy to understand and straightforward to calculate. However, Gower distance is sensitive to non-normality and outliers present in continuous attributes. A pre-processing step for transforming the data might be required in such cases.

2.8   Similarity Functions

An alternative concept to that of the distance is the similarity function $s(\mathbf{x_i}, \mathbf{x_j})$ that compares the two vectors $\mathbf{x_i}$ and $\mathbf{x_j}$ [Duda *et al.* (2012)]. This function should be symmetrical (namely $s(\mathbf{x_i}, \mathbf{x_j}) = s(\mathbf{x_j}, \mathbf{x_i})$ and have a large value when $\mathbf{x_i}$ and $\mathbf{x_j}$ are somehow "similar" and the largest value for identical vectors. A similarity function where the target range is $[0,1]$ is called a dichotomous similarity function.

In fact, the Gower distance metric $d(\mathbf{x_i}, \mathbf{x_j})$ described in the previous section for calculating the "distance" between two instances, can be easily converted to similarity $s(\mathbf{x_i}, \mathbf{x_j})$, using the following transformation:

$$s(\mathbf{x_i}, \mathbf{x_j}) = 1 - d(\mathbf{x_i}, \mathbf{x_j}) \tag{2.14}$$

In the following sub-sections, additional similarity measures are presented.

2.8.1 *Cosine measure*

When the angle between the two vectors is a meaningful measure of their similarity, the normalized inner product may be an appropriate similarity measure:

$$s(\mathbf{x_i}, \mathbf{x_j}) = \frac{\mathbf{x_i}^T \cdot \mathbf{x_j}}{\|\mathbf{x_i}\| \cdot \|\mathbf{x_j}\|} \tag{2.15}$$

2.8.2 *Pearson correlation measure*

The normalized Pearson correlation is defined as:

$$s(\mathbf{x_i}, \mathbf{x_j}) = \frac{(\mathbf{x_i} - \bar{\mathbf{x}}_i)^T \cdot (\mathbf{x_j} - \bar{\mathbf{x}}_j)}{\|\mathbf{x_i} - \bar{\mathbf{x}}_i\| \cdot \|\mathbf{x_j} - \bar{\mathbf{x}}_j\|} \tag{2.16}$$

where $\bar{\mathbf{x}}_i$ denotes the average attribute value over all dimensions.

2.8.3 *Extended Jaccard measure*

The extended Jaccard measure was presented by [Strehl and Ghosh (2000)], and it is defined as:

$$s(\mathbf{x_i}, \mathbf{x_j}) = \frac{\mathbf{x_i}^T \cdot \mathbf{x_j}}{\|\mathbf{x_i}\|^2 + \|\mathbf{x_j}\|^2 - \mathbf{x_i}^T \cdot \mathbf{x_j}} \tag{2.17}$$

2.8.4 *Dice coefficient measure*

The dice coefficient measure is similar to the extended Jaccard measure and it is defined as:

$$s(\mathbf{x_i}, \mathbf{x_j}) = \frac{2\mathbf{x_i}^T \cdot \mathbf{x_j}}{\|\mathbf{x_i}\|^2 + \|\mathbf{x_j}\|^2} \tag{2.18}$$

2.9   Calculating the Dissimilarity Matrix in R

In R, the dissimilarity matrix is calculated using the function *daisy*, which is part of the *cluster* package. The function gets the following parameters:

- x (mandatory): The input dataset — it can be represented as a numeric matrix or data frame. The dissimilarities will be computed among the rows of x.
- Metric (optional, default is Euclidean distance): A string that specifies which metric distance function to use. The supported metrics are: "Euclidian", "manhattan", and "Gower" — all of which were described in the previous sections.
- Stand (optional): A logical flag that indicates if the objects' values should be standardized before calculating the distances.
- Type (optional): A list that specifies the types of the attributes (columns) in the dataset. The supported attribute types are: "symm" (symmetric binary attributes), "asymm" (for asymmetric binary attributes), "ordratio" (ratio scaled values that are treated like ordinary attributes) and "logicalratio" (ratio scaled values that should be logarithmically transformed).

The output of the *daisy* function is the *dissimilarity.object* which represents the dissimilarity matrix of **X**. This matrix is symmetric and hollow because *daisy* uses only metric distance measures. Thus, in order to save storage space, the *dissimilarity.object* stores only the lower triangle (column-wise) as a vector.

The code 2.1 illustrates the dissimilarity matrix calculation for the *flower* dataset which is provided within the *cluster* package. The *flower* dataset contains 18 popular flowers. Each flower is characterized based on the following eight attributes:

(1) Winter — a binary attribute that indicates if the flower can be left outdoors when it freezes.
(2) Shadow — a binary attribute indicating if the flower should be in the shadow.
(3) Tubers — an asymmetric binary attribute that indicate if tubers reproduce the flower.
(4) Color — a nominal attribute that specifies the flower's color (1 = white, 2 = yellow, 3 = pink, 4 = red, 5 = blue).

(5) Soil — an ordinal attribute that indicates the best soil type (1 = dry, 2 = normal , 3 = wet).
(6) Preference — an ordinal attribute containing the user ranking (1 to 18).
(7) Height — a numeric scaled attribute that indicates the flower's height in centimeters.
(8) Distance — a numeric scaled attribute that specifies the distance in centimeters that should be left between any two flowers.

In line 3 of code 2.1, the *daisy* function is executed using the Gower similarity measure (see Section 2.7) that can address objects that are characterized by mixed attributes (numeric, binary and ordinary). The *type* parameters indicate that the first two attributes should be treated as symmetric binary variables, the third binary attribute should be referred to as an asymmetric variable, and the 7th attribute should be treated as an ordinary attribute. In line 4, the summary of the dissimilarity matrix is presented, and in line 5, the *dissimilarity.object* is converted to a full square matrix.

Listing 2.1: Using *daisy* function to calculate the dissimilarity matrix for the flowers dataset

```
1 library(cluster)
2 data(flower)
3 DisMat <- daisy(flower, metric = "gower", stand = FALSE,
      type = list(symm = c(1,2),asymm = 3, ordratio = 7))
4 summary (DisMat)
5 as.matrix(DisMat)
```

Chapter 3

Partitioning Methods for Minimizing Distance Measures

3.1 Introduction

Partitioning methods work by reassigning instances from one cluster to another, starting with an initial partition. These methods typically require the user to specify the number of clusters in advance. Achieving global optimality in partition-based clustering would necessitate exhaustively evaluating all possible partitions, which is impractical. Instead, we use greedy heuristics in the form of iterative optimization. Specifically, a relocation method iteratively shifts points among the k clusters to refine the clustering.

Recall that hard partitioning is defined as a set of clusters $C = \{C_1, C_2, \ldots, C_k\}$ such that $\mathbf{X} = \bigcup_{i=1}^{k} C_i$ and $C_i \bigcap C_j = \emptyset; ; \forall i \neq j$ where k is the number of clusters. This means that every entity belongs to exactly one cluster.

As indicated in Chapter 1, many different types of partitioning methods exist. In this chapter, we mainly focus on center-based methods. The core idea is to identify a clustering structure that minimizes an error criterion measuring the "distance" between each instance and its representative value. The most well-known of these criteria is the Sum of Squared Error (SSE), which calculates the total squared Euclidean distance of instances from their representative values. Achieving a global optimum for SSE would involve exhaustively evaluating every possible partition, but this is highly time-consuming. In fact, finding precisely the best centers is known to be NP-hard, even with just two clusters. A more practical approach is to search for an approximate solution (not necessarily leading to a global minimum) using heuristics. The latter approach is further examined in the following subsections.

3.2   K-Means

3.2.1   *Algorithm overview*

The most straightforward and most commonly used algorithm, which uses a squared error criterion, is the k-means algorithm. This algorithm was first introduced over half a century ago [Lloyd (1982)], but it is still considered one of the most popular data processing algorithms. Moreover, many extensions for the k-means algorithm have been developed since it was first introduced.

K-means algorithm partitions the data into k clusters $(C_1, C_2, \ldots, C_k)$, represented by their centers or means. The center of each cluster is calculated as the mean of all the points belonging to that cluster.

Figure 3.1 presents the pseudo-code of the k-means algorithm. The algorithm starts with an initial set of cluster centers, chosen randomly or according to some heuristic method. In each iteration, every point is assigned to the nearest cluster center based on the Euclidean distance between them. Then, the cluster centers are re-calculated.

The center of each cluster is calculated as the mean of all the instances belonging to that cluster:

$$\mu_k = \frac{1}{N_k} \sum_{q=1}^{N_k} \mathbf{x_q}$$

where N_k is the number of instances belonging to cluster k and μ_k is the mean of the cluster k.

Several convergence conditions are possible. For example, the search may stop when the relocation of the centers does not reduce the partitioning error. This indicates that the present partition is locally optimal. Other stopping criteria can be used also, such as exceeding a pre-defined number of iterations.

Require: X (instance set), k (number of clusters)
Ensure: clusters
1: Initialize k cluster centers.
2: **while** termination condition is not satisfied **do**
3:    Assign instances to the closest cluster center.
4:    Update cluster centers based on the assignment.
5: **end while**

Fig. 3.1: K-means algorithm.

3.2.2 Illustration of k-means algorithm

We would like to illustrate the k-means algorithm using a subset of the Iris dataset. Table 3.1 presents this subset, which consists of 40 instances

Table 3.1: Iris dataset.

Sepal Length	Sepal Width
5.4	3.4
4.8	3.4
5.8	2.7
6	2.2
5.7	2.5
6.9	3.1
5.9	3.2
6.4	3.2
5.8	2.7
4.6	3.1
4.7	3.2
6.3	2.5
6.7	3
6.7	3
5.7	3
6.4	2.7
5.5	4.2
4.4	2.9
7.3	2.9
6.2	2.9
5	3.2
6	3
5.6	3
5.4	3.7
7.9	3.8
6.3	2.5
6.1	3
5	3.4
4.8	3.1
5.4	3.9
6.3	2.8
4.9	2.4
5.1	3.8
5.5	2.5
6.2	2.8
6.7	3.1
6.3	2.7
6.3	3.4
5	3.6
5	3.4

that were randomly sampled from the original Iris dataset. For the sake of clarity, we are using only two features (Sepal Length and Sepal Width) out of the four input features that the original Iris dataset includes.

In this illustration, we would like to group the instances into three clusters (i.e., k=3). We begin by randomly selecting three instances from the subset that will be used as the initial centroids. Table 3.2 shows these three selected instances. These are instances 15, 38 and 39 of Table 3.1.

Figure 3.2 visualizes the dataset of Table 3.1. Note that the three centroids are displayed with a unique marker shape (circle, rectangle and rhombus) and a solid fill.

We begin with the first iteration of the k-means algorithm. We calculate the Euclidian distance of each instance to each one of the Centroids. Recall that the Euclidian distance of an instance $\mathbf{x_i} \in \mathbb{R}^n$ to a centorid $\mathbf{c_j}$ is calculated as following:

$$D\left(\mathbf{x_i}, \mathbf{c_j}\right) = \sqrt{\left(x_{i,1} - c_{i,1}\right)^2 + \left(x_{i,2} - c_{i,2}\right)^2 + \cdots + \left(x_{i,n} - c_{i,n}\right)^2} \quad (3.1)$$

Table 3.2: Initial centroids for the Iris dataset.

Centroid Number	Sepal Length	Sepal Width
1	5.7	3
2	6.3	3.4
3	5	3.6

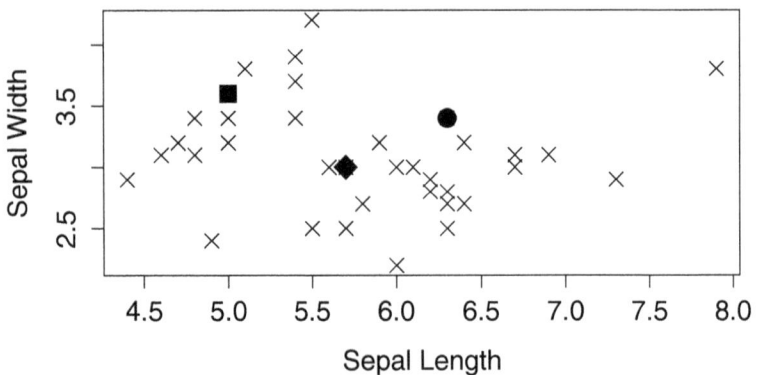

Fig. 3.2: Visualization of the Iris dataset.

Specifically, in the current example, the Euclidian distance results in the following equation:

$$D\left(\mathbf{x_i}, \mathbf{c_j}\right)$$

$$= \sqrt{\left(x_{i,\text{Sepal Length}} - c_{i,\text{Sepal Length}}\right)^2 + \left(x_{i,\text{Sepal Width}} - c_{i,\text{Sepal Width}}\right)^2} \tag{3.2}$$

For example the Euclidean distance of instance $\mathbf{x_1} = \left(5.4\ 3.4\right)$ to $\mathbf{c_1} = \left(5.7\ 3\right)$ is:

$$D\left(\mathbf{x_1}, \mathbf{c_1}\right) = \sqrt{\left(5.4 - 5.7\right)^2 + \left(3.4 - 3\right)^2} = 0.5 \tag{3.3}$$

Table 3.3 presents the outcome of this calculation.

Next, we assign each instance to its nearest centroid, as illustrated in Table 3.4. This is known as the assignment step.

Following the assignment step, we perform the update step, in which we update the centroid coordinates. The centroid of cluster j is the arithmetic mean ("average") position of all the instances that are assigned to this cluster. Table 3.5 presents the new centroid.

Figure 3.3 shows the new centroid and the assignment of each instance to the nearest centroid. The marker shape of each instance indicates to which cluster (centroid) it was assigned. Note that the instances' shape has no fill while the centroids' shapes are solid fill.

Figures 3.4 and 3.7 illustrate the subsequent iterations. Figure 3.6 show all iterations in the same chart. The initial centroids are filled in with a light grey. After the first iteration, the obtained centroids are filled in with a darker grey and so forth. The algorithm continues until the assignments no longer change.

3.2.3   *Running k-means in R*

In the first chapter, we have already shown how to run the k-means algorithm using the implementation that is provided as part of the basic *stats* package. Other implementations of k-means can be found in several other packages. In this section, we will illustrate the usage of *amap* (short for Another Multidimensional Analysis Package) for this purpose. Because the *amap* package (as opposed to the *stats* package) is not automatically loaded at the start of an R session, we need to install it and then load it before we use it for the first time. If you are using R-Studio, installation of a package is simply done by selecting *Tools* in the menu bar and then

Table 3.3: Calculating the distance to the centroids in the Iris dataset.

Sepal Length	Sepal Width	Distance Centroid 1	Distance Centroid 2	Distance Centroid 3
5.4	3.4	0.50	0.90	0.45
4.8	3.4	0.98	1.50	0.28
5.8	2.7	0.32	0.86	1.20
6	2.2	0.85	1.24	1.72
5.7	2.5	0.50	1.08	1.30
6.9	3.1	1.20	0.67	1.96
5.9	3.2	0.28	0.45	0.98
6.4	3.2	0.73	0.22	1.46
5.8	2.7	0.32	0.86	1.20
4.6	3.1	1.10	1.73	0.64
4.7	3.2	1.02	1.61	0.50
6.3	2.5	0.78	0.90	1.70
6.7	3	1.00	0.57	1.80
6.7	3	1.00	0.57	1.80
5.7	3	0.00	0.72	0.92
6.4	2.7	0.76	0.71	1.66
5.5	4.2	1.22	1.13	0.78
4.4	2.9	1.30	1.96	0.92
7.3	2.9	1.60	1.12	2.40
6.2	2.9	0.51	0.51	1.39
5	3.2	0.73	1.32	0.40
6	3	0.30	0.50	1.17
5.6	3	0.10	0.81	0.85
5.4	3.7	0.76	0.95	0.41
7.9	3.8	2.34	1.65	2.91
6.3	2.5	0.78	0.90	1.70
6.1	3	0.40	0.45	1.25
5	3.4	0.81	1.30	0.20
4.8	3.1	0.91	1.53	0.54
5.4	3.9	0.95	1.03	0.50
6.3	2.8	0.63	0.60	1.53
4.9	2.4	1.00	1.72	1.20
5.1	3.8	1.00	1.26	0.22
5.5	2.5	0.54	1.20	1.21
6.2	2.8	0.54	0.61	1.44
6.7	3.1	1.00	0.50	1.77
6.3	2.7	0.67	0.70	1.58
6.3	3.4	0.72	0.00	1.32
5	3.6	0.92	1.32	0.00
5	3.4	0.81	1.30	0.20

selecting *Install Packages* in the sub-menu and finally entering the name of the package to be installed (*amap* in this case). Alternatively, you can install a package using the shell command *install.packages ("amap")*.

Once installed, the package can be loaded using the library command as illustrated in the code 3.1. In line 2, we create a two-dimensional

Table 3.4: Assignment of instances to the nearest centroids in the Iris dataset.

Sepal Len.	Sepal W.	Distance Cen. 1	Distance Cen. 2	Distance Cen. 3	Cluster
5.4	3.4	0.50	0.90	0.45	3
4.8	3.4	0.98	1.50	0.28	3
5.8	2.7	0.32	0.86	1.20	1
6	2.2	0.85	1.24	1.72	1
5.7	2.5	0.50	1.08	1.30	1
6.9	3.1	1.20	0.67	1.96	2
5.9	3.2	0.28	0.45	0.98	1
6.4	3.2	0.73	0.22	1.46	2
5.8	2.7	0.32	0.86	1.20	1
4.6	3.1	1.10	1.73	0.64	3
4.7	3.2	1.02	1.61	0.50	3
6.3	2.5	0.78	0.90	1.70	1
6.7	3	1.00	0.57	1.80	2
6.7	3	1.00	0.57	1.80	2
5.7	3	0.00	0.72	0.92	1
6.4	2.7	0.76	0.71	1.66	2
5.5	4.2	1.22	1.13	0.78	3
4.4	2.9	1.30	1.96	0.92	3
7.3	2.9	1.60	1.12	2.40	2
6.2	2.9	0.51	0.51	1.39	2
5	3.2	0.73	1.32	0.40	3
6	3	0.30	0.50	1.17	1
5.6	3	0.10	0.81	0.85	1
5.4	3.7	0.76	0.95	0.41	3
7.9	3.8	2.34	1.65	2.91	2
6.3	2.5	0.78	0.90	1.70	1
6.1	3	0.40	0.45	1.25	1
5	3.4	0.81	1.30	0.20	3
4.8	3.1	0.91	1.53	0.54	3
5.4	3.9	0.95	1.03	0.50	3
6.3	2.8	0.63	0.60	1.53	2
4.9	2.4	1.00	1.72	1.20	1
5.1	3.8	1.00	1.26	0.22	3
5.5	2.5	0.54	1.20	1.21	1
6.2	2.8	0.54	0.61	1.44	1
6.7	3.1	1.00	0.50	1.77	2
6.3	2.7	0.67	0.70	1.58	1
6.3	3.4	0.72	0.00	1.32	2
5	3.6	0.92	1.32	0.00	3
5	3.4	0.81	1.30	0.20	3

Table 3.5: Recalculating the centroids in the Iris dataset.

Centroid Number	Sepal Length	Sepal Width
1	6.709	3.082
2	5.007	3.45
3	5.873	2.713

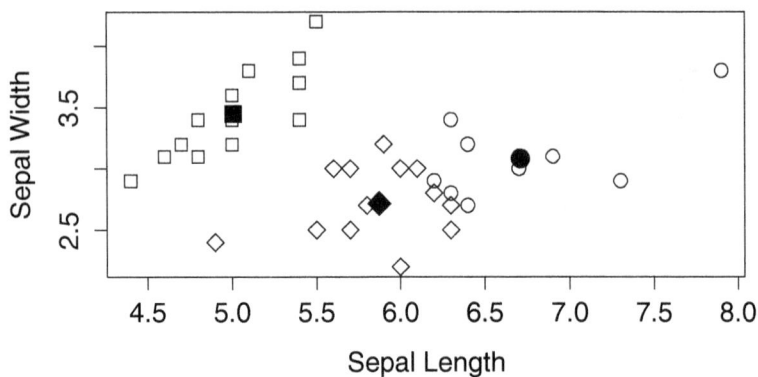

Fig. 3.3: Visualization of the Iris dataset assignment to clusters after one iteration.

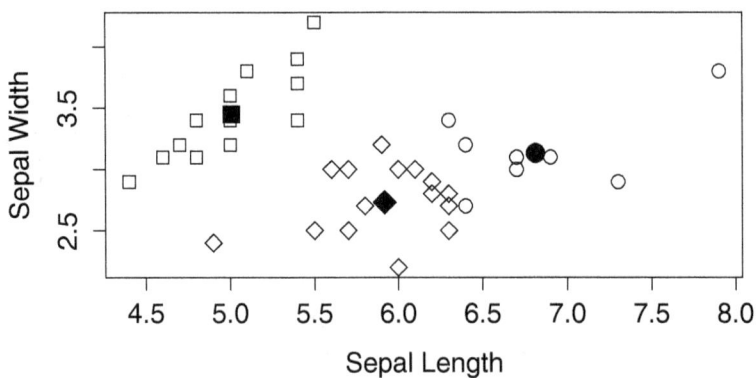

Fig. 3.4: Visualization of the Iris dataset assignment to clusters after two iterations.

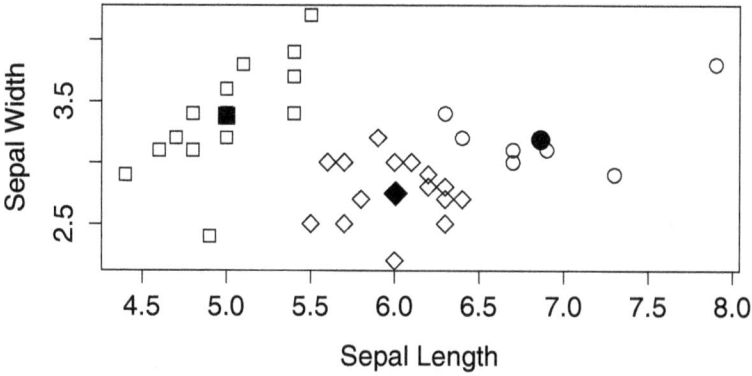

Fig. 3.5: Visualization of the Iris dataset assignment to clusters after three iterations.

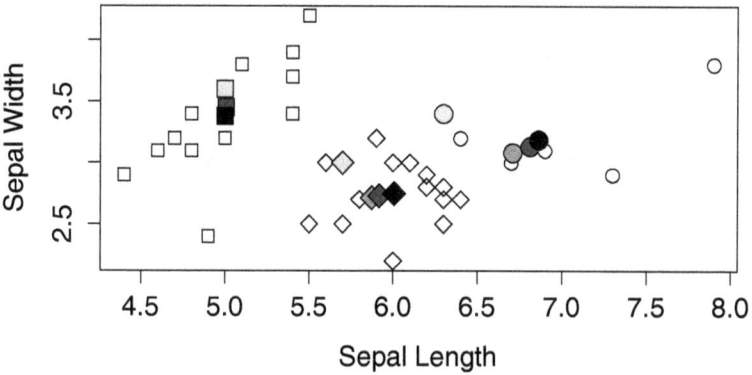

Fig. 3.6: Visualization of the final clustering of Iris dataset.

dataset that consists of 40 instances that were sampled from the original Iris dataset. In line 3, the Kmeans algorithm is run with the following setting: the number of clusters is 3 (*centers = 3*); The distance metric is Euclidean (*method = "euclidean"*) and the maximum number of iterations that the algorithm can perform is 100 (*iter.max = 100*). The returned clustering object is assigned to the variable *cl*. In line 4, the dataset used for the clustering is plotted, and in line 5, we add to the graph the centers of the obtained three clusters.

```
1 library(amap)
2 iris2dsample <- iris[sample(nrow(iris),40),1:2]
3 cl <- Kmeans(iris2dsample, centers=3, method="euclidean",
     iter.max = 100)
4 plot(iris2dsample,cex=1.5,pch=4,xlab="Sepal Length", ylab="
     Sepal Width")
5 points(cl$centers, pch = 21:23, cex=2,bg="grey1")
```

In addition to Euclidean distance, other distance metrics can be applied. Given two vector instances x_i and x_j, the following distances metrics are defined:

(1) Manhattan: Absolute distance between the two vectors (i.e. l-1 norm) as defined in Section 2.4.

(2) Correlation: This measure refers to the centered Pearson correlation as defined in Section 2.8.2.

(3) Abscorrelation: The absolute value of the centered Pearson measure.

(4) Pearson: This measure refers to the not-centered Pearson measure.

(5) Abspearson: The absolute value of the not centered Pearson measure.

(6) Maximum: the distance between two vectors is the greatest of their differences along any coordinate dimension. As indicated in Section 2.4 this is usually referred to as Chebychev distance.

(7) Canberra: The distance in this case is defined as:

$$D_{\text{Canberra}}\left(x_i, x_j\right) = \sum_{l=1}^{n} \frac{|x_{i,l} - x_{j,l}|}{|x_{i,l} + x_{j,l}|} \tag{3.4}$$

where the numerator or denominator with zero values is omitted from the sum.

(8) Binary: The vectors are regarded as binary vectors, such that non-zero elements are considered as true, and zero elements are considered as false. The distance is defined as the proportion of entries in which only one of the vectors is true amongst all entries in which at least one of the vectors is true.

(9) Spearman: Spearman's rank-order correlation coefficient is defined as the distance among the ranks of the vectors' elements:

$$D_{\text{Spearman}}\left(x_i, x_j\right) = \sum_{l=1}^{n} \left(r_{i,l} - r_{j,l}\right)^2 \tag{3.5}$$

where $r_{i,l}$ and $r_{j,l}$ denote the rank of element l in the vectors x_i and x_j.

Spearman rank distance can overcome the drawback of Pearson's metric that assumes the features values are distributed according to the Gaussian distribution and is not useful when this assumption is violated. Spearman rank replaces the feature values by their corresponding ranks and therefore can be used when the data is not Gaussian distributed. That being said, the conversation of the numerical values to their ranks results in a loss of information, thus Pearson's correlation coefficient usually outperforms Spearman rank.

(10) Kendall: Kendall's measure is also based on the values' ranks. Specifically it is defined as:

$$D_{\text{Kendall}}\left(\mathbf{x_i}, \mathbf{x_j}\right) = \sum_{k,l} I\left(r_{i,l}, r_{i,k}, r_{j,l}, r_{j,k}\right) \tag{3.6}$$

where I is a sign indicator that gets the value 0 if $x_{i,l}$, $x_{i,k}$ in same order as $x_{j,l}$, $x_{j,k}$, 1 if not. More specifically, the function I is defined as:

$$I\left(r_{i,l}, r_{i,k}, r_{j,l}, r_{j,k}\right)$$
$$= \begin{cases} 0 & (r_{i,l} - r_{i,k})(r_{j,l} - r_{j,k}) > 0 \; or \; r_{i,l} - r_{i,k} = r_{j,l} - r_{j,k} = 0 \\ 1 & else \end{cases}$$

$$\tag{3.7}$$

3.2.4 The properties of k-means algorithm

K-means algorithm is a gradient-decent procedure, which begins with an initial set of k cluster-centers and iteratively updates it so as to decrease the error function. As indicated above, k-means is a very popular clustering algorithm and, in many cases, is used as a method of choice, particularly when the data instances can be clustered into groups of hyper-spherical shape.

The complexity of T iterations of the k-means algorithm performed on a sample size of m instances, each characterized by n attributes, is: $O(T * k * m * n)$. A rigorous proof of the finite convergence of the k-means type algorithms is given in [Selim and Ismail (1984)]. Since the required number of iterations T is negligible in most cases and both k and n are usually much less than m, the time complexity becomes practically linear. This linear complexity is one of the reasons for the popularity of the k-means algorithm. Even if the number of instances is substantially large (which often is the case nowadays), this algorithm is computationally attractive. In particular, the k-means algorithm has an advantage over other clustering methods (e.g., hierarchical clustering methods), which have non-linear complexity. Thus, k-means is a good choice for clustering large-scale data sets. Other reasons for the algorithm's popularity are its ease of interpretation, simplicity of implementation, speed of convergence and adaptability to sparse data [Dhillon and Modha (2001)]. The Achilles heel of the k-means algorithm involves the selection of the initial partition. The selection of the initial partition significantly influences the algorithm, potentially leading to either a global or local minimum.

As a typical partitioning algorithm, the k-means algorithm works well only on data sets with isotropic clusters, and is less versatile than single link algorithms, for instance. In addition, this algorithm is sensitive to noisy data and outliers (a single outlier can increase the squared error dramatically); it is applicable only when the mean is defined (namely, for numeric attributes); and it requires the number of clusters in advance, which is not trivial when no prior knowledge is available.

3.3   Determining the Number of Clusters

The most important parameter in k-means is k, which is the assumed number of clusters in the dataset. The need to specify the value of k is considered to be one of the disadvantages of the k-means algorithm. One way to determine the number of clusters in a dataet, is to simply try different values of k and compare their performance according to some validity measure. Given the performance of the various values of k, we can either choose the value of k with the best performance, or apply the elbow method. The main idea of the elbow method is to choose a number of clusters such that increasing this number does not significantly improve the evaluation measure. Various validity measures have been suggested in the literature. In this section, we will discuss only a few of them. A much more profound survey is given in Chapter 6.

A well-known evaluation measure is the percentage of variance explained, which is defined as the ratio of the between-cluster sum of squares of deviations to the total sum of squares of deviations. Specifically, within-cluster sum of square (or withinSS for short) is defined as:

$$withinSS = \sum_{j=1}^{k} \sum_{\mathbf{x_i} \in C_j} \left((x_{i,1} - c_{j,1})^2 + (x_{i,2} - c_{j,2})^2 + \cdots + (x_{i,n} - c_{j,n})^2 \right)$$

$$(3.8)$$

where the notation $\mathbf{x_i} \in C_j$ indicates that cluster C_j is the nearest center to instance $\mathbf{x_i}$.

The total sum of the squares is the **sum** of the squared deviations of the instances about their mean. It is defined as:

$$totalSS = \sum_{i=1}^{m} \left((x_{i,1} - \hat{x}_{\cdot,1})^2 + (x_{i,2} - \hat{x}_{\cdot,2})^2 + \cdots + (x_{i,n} - \hat{x}_{\cdot,n})^2 \right) \quad (3.9)$$

where $\hat{x}_{\cdot,j}$ is the mean value of the j-th dimension. Finally, the between-cluster sum of square is defined as:

$$betweenSS = totalSS - withinSS \quad (3.10)$$

Code 3.2 illustrates the process of selecting the number of clusters based on the percentage of variance explained. In lines 1-2, we create a copy of the Iris dataset while removing the class attribute (the Species attribute is set to null because it is not used in the clustering). In lines 5-8, we try different values of k between 2 to 30, and for each one of the value of k, the percentage of variance explained is calculated by dividing the between-cluster sum of square by the total sum of the square. Note that the clustering object includes the components *betweenss* and *totss*, which hold the between-cluster sum of squares and the total sum of squares, respectively. The program plots the percentage of variance explained by the clusters against the number of clusters. While the first clusters explain a lot of the variance, at some point, the marginal gain becomes relatively small and insignificant, and one can notice a bend in the graph. Thus, the number of clusters is chosen at this point.

Listing 3.2: Finding the optimal number of cluster in Iris dataset

```
1  Cleariris <- iris
2  Cleariris$Species <- NULL
3
4  perc=array(,30)
5  for (i in 2:30) {
6     cl = kmeans(Cleariris, centers=i, iter.max = 100)
7     perc[i]<-cl$betweenss/cl$totss
8  }
9
10 plot(1:30, perc, type="b", xlab="Number of Clusters",
11      ylab="Within groups sum of squares")
```

Ray and Turi [Ray and Turi (1999)] introduced the compactness-separation measure for selecting the optimal number of clusters. Given k clusters, $C_1, \ldots, C_k$ and the corresponding coordinates of their centers

$\mathbf{c_1}, \ldots, \mathbf{c_k}$ the intra-cluster distance measure M_{intra} is defined as:

$$M_{intra} = \frac{withinSS}{m} \tag{3.11}$$

where m is the number of instances in the dataset. The inter-cluster measure M_{inter} is defined as:

$$M_{inter} = \min_{1 \leq i < j \leq k} \left((c_{i,1} - c_{j,1})^2 + (c_{i,2} - c_{j,2})^2 + \cdots + (c_{i,n} - c_{j,n})^2 \right) \tag{3.12}$$

The compactness-separation validity measure is simply defined as, where lower values considered to be a better clustering:

$$M_{CS} = \frac{M_{intra}}{M_{inter}} \tag{3.13}$$

3.3.1   *The clues package*

Clues package aims to provide functions for automatically estimating the number of clusters and getting the final cluster [Chang *et al.* (2010)]. The CLUES (CLUstEring based on local Shrinking) algorithm consists of three phases [Wang *et al.* (2007)]: A) shrinking in which data points are pushed towards a denser region; B) partitioning in which the shrunk data is decomposed into clusters; C) and finally, the algorithm determines the optimal k based on either the CH index or the Silhouette index. The CH index is the ratio of *betweenSS* to *withinSS* formulated in Equations (3.10) and (3.8), respectively. The Silhouette index measures the average distance between an object in a cluster to all the objects in the nearest neighboring cluster of this object. The search for the optimal k begins with a small number of clusters and gradually increases it until a sufficient index value is reached.

List 3.3 illustrates CLUES implementation on the Iris dataset. Line 3 defines the dataset for clustering as having two columns (1,4) in Iris dataset. In line 4, the CLUES algorithm is applied, and its results are stored in the variable *ClusteringResults*. In line 5, the results are printed, including the optimal number of clusters and the Silhouette index value obtained for this optimal setting.

Listing 3.3: Using CLUES for estimating the number of clusters

```
1 library("clues")
2
3 x=cbind(iris[,1],iris[,4])
4 ClusteringResults <- clues(x, quiet = TRUE)
5 # get summary statistics
6 summary(ClusteringResults)
```

3.4 X-Means

3.4.1 *Algorithm overview*

In the previous section, we have seen several methods for estimating the number of clusters. In this section, the x-means algorithm is presented, which, among other advantages, tries to automatically estimate the correct number of clusters. X-means is an extension of the popular k-means algorithm that addresses some of its shortcomings [Pelleg *et al.* (2000)]. Specifically, it aims to improve the computational scalability of k-means, efficiently and automatically search for the best number of clusters and attempt to avoid local minima. The main idea is to use k-means with k=2 to perform the initial clustering and then continue to partition the clusters into smaller clusters. The algorithm has two primary operations: Improve-Params, which aims to run conventional k-means, and Improve-Structure, which determines if the current cluster should be further split into subclusters. In order to decide if and which cluster to partition, the algorithm uses the Bayesian information criterion (BIC) [Schwarz *et al.* (1978)] or Akaike information criterion (AIC)[Akaike (1998)].

Given a set of different k-means clustering $\mathbf{C}$ each of which was obtained using a different value of k. Bayesian information criterion can be used to choose the best clustering based on the following formula:

$$BIC(\mathbf{C}^{(i)}) = I_j(\mathbf{X}) - \frac{k_j \cdot n \cdot log(m)}{2}$$

where m is the dataset size (i.e. number of instances in $\mathbf{X}$), n is the number of features and k_j is the number of clusters in $\mathbf{C}^{(i)}$. The function $I_j(\mathbf{X})$ represents the log-likelihood of the dataset assuming identical spherical Gaussian distribution with the parameters obtained according to the clustering $\mathbf{C}^{(i)}$. Instead of BIC, other common information criteria such as AIC and MDL (short for Minimum Description Length) can be used. The reason BIC is chosen over other criteria is that BIC relies on prior probability rather than the distance between two distributions. As mentioned above x-means contains other enhancements that aim to improve speed. In particular, X-means embeds the dataset in a multiresolution kd-tree and stores sufficient statistics at its nodes [Pelleg *et al.* (2000)]. These statistics can be used to make a local decision that maximizes the clustering's BIC values.

3.4.2   Running X-means in R

While there is no native implementation of X-means in CRAN (The Comprehensive R Archive Network), it is still possible to run X-means in R using a special interface to Weka. Weka is a popular machine learning suit developed at the University of Waikato. Weka is an open-source software written in Java that is available under the GNU General Public License. *RWeka* package contains the interface from R to Weka and the Weka jar is included in a separate package called *RWekajars*. Thus, the package *RWeka* should be installed. This will also install the required *RWekajars* package.

Starting from version 3.7, Weka is installed with only the most popular learning algorithms. In order to install other algorithms, one should use the package manager to select which additional component is to be installed. This is done in R using *WPM* function. Code 3.4 illustrates the process. In the first, the *RWeka* package is loaded. In the second line, the *XMeans* algorithm is added to Weka. In lines 3-4, we create a copy of the Iris dataset while removing the class attribute (the Species attribute is set to null because it is not used in the clustering). In line 6, we call *XMeans* with the following parameters:

(1) "-H", 12 — which indicates that the maximum number of clusters is 12.
(2) "-use-kdtree" which indicates that KDTree data structure should be used internally.
(3) "-K", "weka.core.neighboursearch.KDTree -P" which indicate the full class name of KDTree class to use, followed by scheme options.

After running the code, the last line presents the resulting clustering, which includes, in this case, two clusters that have obtained a BIC value of 89.61263.

```
Listing 3.4:  Clustering Iris dataset using X-means
1 library ("RWeka")
2 WPM("install-package", "XMeans")
3 Cleariris <- iris
4 Cleariris$Species <- NULL
5
6 cl2 <- XMeans(Cleariris, c("-H", 12, "-use-kdtree", "-K", "
      weka.core.neighboursearch.KDTree -P"))
7 print(cl2)
```

Table 3.6: XMeans list of parameters.

Parameter name	Description	Default value
I	maximum number of overall iterations	1
M	maximum number of iterations in the kMeans loop in the Improve-Parameter part	1000
J	maximum number of iterations in the kMeans loop for the splitted centroids in the Improve-Structure part	1000
L	minimum number of clusters	2
H	maximum number of clusters	4
B	distance value for binary attributes	1.0
use-kdtree	Uses the KDTree internally	no
K	Full class name of KDTree class to use, followed by scheme options.	no KDTree class used
C	cutoff factor, takes the given percentage of the splitted centroids if none of the children win	0.0
D	Full class name of Distance function class to use, followed by scheme options.	EuclideanDistance
N	file to read starting centers from	no file
O	file to write centers to	no file
U	The debug level	0
Y	The debug vectors file	no file
S	Random number seed	10

In the above example, we have set some parameters controlling XMeans. Because these parameters are the internal parameters of the Weka's implementation, they should be provided within a *Weka control* object. This is the reason why the list of parameters is not presented in R help when the command *help(XMeans)* is executed. Thus, for the sake of clarity, the complete list of parameters is provided in Table 3.6.

3.5 K-Means++

3.5.1 *Algorithm overview*

K-means is considered to be a simple approximation of the optimal partitioning. However, the obtained approximation is not guaranteed to be a good one. In fact, the accuracy of the approximation depends on the initial k arbitrary centers that are randomly selected.

K-means++ suggests a better way to initialize k-means by choosing random starting centers with specific probabilities [Arthur and Vassilvitskii (2007)]. Specifically the first center c_1 is selected uniformly at random from

X similarly to how the original k-means select all centers. However, the subsequent centers $(\mathbf{c_1}, \ldots, \mathbf{c_k})$ are chosen sequentially according to the probability:

$$p(x) = \frac{D(x)^2}{\sum_{x \in X} D(x)^2} \qquad (3.14)$$

where $D(\mathbf{x})$ denotes the shortest distance of a data point $\mathbf{x}$ to the closest center we have already chosen in previous iterations. The exact k-means++ algorithm is described in Figure 3.7.

The k-means++ method for selecting the initial centers significantly improves the accuracy of the original k-means. The extra time required to select the initial centers is negligible because once the new centers are selected using the k-means++ procedure, the clustering part itself converges much faster than the original k-means algorithm. Various experiments have shown that k-means++ substantially outperforms standard k-means in terms of both accuracy and speed. Specifically, the inventors of k-means++ reported on a typical 2-fold improvements in computation cost, and for certain datasets, they reported close to 1000-fold improvements in accuracy. Moreover, while the clusterings of the original k-means can be arbitrarily worse than the optimum, the k-means++ algorithm guarantees an approximation ratio that is $O(logk)$-competitive with the optimal clustering.

Require: X (instance set), k (number of cluster)
Ensure: clusters
1: Choose first center c_1 uniformly at random from X. .
2: **for** i=2 to k **do**
3:     **for all** xinX **do**
4:         Find distance to the nearest center: $D(\mathbf{x}) = \min_{j \in 1, \cdot, i-1} D(c_j, \mathbf{x})$.
5:     **end for**
6:     Choose a new data point at random as a new center c_i, using a weighted probability distribution presented in Equation (3.14).
7: **end for**
8: Proceed using standard k-means.

Fig. 3.7: K-means++ algorithm.

3.5.2 Running k-means++ in R

There are several implementations of k-means++ in R, including function *kcca* (in package *flexclust*) and function *kmeanspp* (in *LICORS* package). The Code 3.5 demonstrates how the package *flexclust* can be used to apply k-means++ on Iris dataset. The first line in Code 3.5 loads the package *flexclust* assuming this package was already installed. The second line aims to load the built-in Iris dataset. For the sake of visibility, lines 3 and 4, select a sample of 40 points and project them using only two dimensions (Sepal Length and Sepal Width). In line 5, we cluster the data using the k-means algorithm *(family=kccaFamily("kmeans"))* centers' initialization using k-means++ procedure *(initcent="kmeanspp")*. Finally, in line 6, we plot the obtained clusters, and in line 7, we add the original points to the plots to illustrate their clusters' assignments.

Listing 3.5: Using k-means++ clustering to cluster Iris dataset

```
1 library(flexclust)
2 data(iris)
3 idx <- sample(1:dim(iris)[1], 40)
4 X=iris[idx,1:2]
5 cl = kcca(X, k=2, family=kccaFamily("kmeans"), control=list(
      initcent="kmeanspp"))
6 image(cl,xlab="Sepal Length", ylab="Sepal Width")
7 points(X,cex=1.5,pch=4)
```

3.6 K-Medoids: Partitioning Around Medoids

3.6.1 Algorithm overview

Another partitioning algorithm which attempts to minimize the sum of squares is the k-medoids or PAM (partition around medoids — [Kaufman and Rousseeuw (1987)]). This algorithm is very similar to the k-means algorithm. It differs from the latter mainly in its representation of the different clusters. Each cluster is represented by the most centric entity in the cluster (called medioid), rather than by the implicit mean that may not belong to the cluster. Practically, the medoid is the entity with the minimum distance to the other members of the cluster. Figure 3.8 presents the pseudo-code of k-medoids. Each iteration attempts to swap the existing medoid with a non medoid entity that decreases the total distances.

Require: X (instance set), k (number of clusters)
Ensure: clusters
  1: Initialize k medoids by randomly selecting k different instances from
     X.
  2: $TotalDistance \leftarrow \infty$
  3: **repeat**
  4:    Assign instances to the closest medoid.
  5:    Compute the total distance of all instances to their closest medoid
  6:    **for each** medoid m_i **do**
  7:       **for each** non-medoid instance x_j that is assigned to m_i **do**
  8:          Compute the total distance by swapping x_j with m_i
  9:          **if** Total distance has not decreased **then**
10:            Undo the swap
11:          **end if**
12:       **end for**
13:    **end for**
14: **until** No decrease in the $TotalDistance$

Fig. 3.8: K-Medoids algorithm.

    When dealing with noise and outliers, the k-medoids method is a more robust alternative to the k-means algorithm. This is because a medoid is less affected by outliers or extreme values compared to a mean. However, the computation involved in the k-medoids method is more intensive than that of the k-means method. Both methods necessitate the user to specify the value of k, which denotes the number of clusters.

3.6.2   *Running k-medoids in R*

The k-medoids is implemented in R as part of the *cluster* package. List 3.6 illustrates this implementation. In line 5, the clustering is performed on a two-dimensional sample of the Iris dataset. In addition, to provide the dataset (x) and the number of clusters (k), we set the distance metric to Euclidean and that the values of the features should be standardized before calculating distances (*stand=TRUE*).

Listing 3.6: Clustering Iris dataset with PAM algorithm

```
1 library("cluster")
2 iris2dsample <- iris[sample(nrow(iris),40),1:2]
3
4 #execute pam algorithm with the iris dataset
5 cl <- pam(x=iris2dsample, k=2, stand=TRUE, metric="euclidean
     ")
6
7 #print the obtained clustering
8 summary(cl)
9
10 #plot a graphic showing the clusters and the medoids of each
     cluster
11 plot(iris2dsample, pch = cl$clustering,cex=1)
12 points(cl$medoids, pch = 1:2,cex=1.5)
```

3.6.3 *PROCLUS and ORCLUS algorithms*

Several variations and extensions have been developed for the original k-medoids algorithm. PROCLUS (PROjected CLUSting)[Aggarwal *et al.* (1999)] aims to work efficiently in high dimensional space by picking closely correlated dimensions and finding clusters in the corresponding projected subspace. One unique advantage of PROCLUS is that the subsets of dimensions are selected specifically for each cluster. Once the sparse subspaces of each cluster are eliminated, the instances are projected into the subspaces in which the greatest similarity occurs. The user provides the number of clusters k and the dimensionality of the subspace l as input parameters. The output is a set of clusters together with a possibly different subspace for each cluster.

PROCLUS consists of three phases: the initialization phase, the iteration phase, and the refinement phase. In the initialization phase, an initial set of k medoids is chosen. In the second phase, the algorithm iteratively replaces bad medoids with new ones in order to reduce the total distance. In the refinement phase, each medoid is represented by a new set of dimensions and all entities in the dataset are reassigned to the medoids using the new sets of dimensions.

PROCLUS has been later extended and generalized in another algorithm called ORCLUS (arbitrarily ORiented projected CLUSter generation) [Aggarwal and Yu (2002)], which projects the original dimensions using linear projection instead of simply selecting the most informative dimensions. Code 3.7 illustrates how to run the ORCLUS algorithm in R using *ORCLUS* package. As indicated above, the most important parameters are the number of clusters (k) and the number of dimensions (l). In line 5, the clustering is performed. In line 6 the clustering assignments of all data instances are printed. In line 7 the coordinates of the clusters' centers (in the original space) are printed. In line 8, the linear projections from the original four dimensions to two dimensions for each cluster are printed.

Listing 3.7: Clustering Iris dataset with ORCLUS algorithm

```
1  library("orclus")
2  Cleariris <- iris
3  Cleariris$Species <- NULL
4
5  cl <- orclus(x = Cleariris, k = 3, k0=10, l = 2)
6  print(cl$cluster)
7  print(cl$centers)
8  print(cl$subspaces)
```

3.6.4   *CLARA and CLARANS algorithms*

The original PAM algorithm works inefficiently for a large dataset due to its computational complexity, mainly because it examines all instances that are not medoids in each iteration. Specifically, each iteration of PAM has a computation complexity of $O(k(n-k)^2)$. Clustering LARge Applications (CLARA) [Kaufman and Rousseeuw (2009)], a sampling-based method, aims to deal with large datasets. The idea is to draw multiple subsamples of the dataset, then apply PAM on each subsample, and select the best clustering. The quality of a clustering is based on the total distances of all entities in the dataset. Experimental studies show that five subsamples when each of which consists of $40 + 2 \cdot k$ instances are sufficient to provide good results. Given that the computational complexity of CLARA is only $O(k(40+k)^2 + k(n-k))$ which is lower than the computational complexity of PAM. The algorithm CLARA is implemented in the *cluster* package. List 3.8 illustrates this implementation. In line 5, the clustering is performed on a two-dimensional sample of the Iris dataset. In addition, to providing the dataset (x) and the number of clusters (k), we set the distance metric to Euclidean and that the values of the features should be standardized before calculating distances (*stand=TRUE*). Moreover, we set the number of samples to be drawn from the dataset to 7, and the size of each sample to 30.

Listing 3.8: Clustering Iris dataset with CLARA algorithm

```
1  library("cluster")
2  iris2dsample <- iris[sample(nrow(iris),40),1:2]
3
4  #execute pam algorithm with the iris dataset
5  cl <- pam(x=iris2dsample, k=2, stand=TRUE, metric="euclidean
      ",
6          samples = 7, sampsize =   30)
7
8  #print the obtained clustering
9  summary(cl)
10
11 #plot a graphic showing the clusters and the medoids of each
      cluster
12 plot(iris2dsample, pch = cl$clustering,cex=1)
13 points(cl$medoids, pch = 1:2,cex=1.5)
```

CLARANS (Clustering Large Applications based on RANdomized Search) algorithm improves CLARA by not confining itself to any sub-sample at any given time. Practically CLARA algorithm can be seen as a graph-searching algorithm. Each node in the graph represents an optional clustering solution (i.e. a different set of k medoids). The entire graph represents the search space. Nodes in the graph are considered neighbors if they differ in one medoid. Instead of searching the entire graph, CLARA reduces the time complexity by selecting a sample of the data and confining the search to the subgraph induced by this sample. CLARA selects the subsamples a priori. Thus, the search space is localized to a specific subsample and it might overlook some good candidate nodes in the original complete graph. CLARANS, on the other hand, considers the entire graph as in PAM. However, instead of examining all neighbors, CLARANS draws a sample of neighbors in each iteration and evaluates only this sample. Experimental studies show that CLARANS is particularly useful in identifying spatial structures that may be present in the dataset. CLARANS is a few times faster than PAM and provides better quality, and it creates clusterings that have better quality than those produced by CLARA.

Park and Jun [Park and Jun (2009)] present another fast implementation for K-medoids clustering. It reduces the time complexity by calculating the distance matrix once and using it to find new medoids at every iterative step. The algorithm consists of three phases. In the first phase, the initial set of medoids is selected. In the second phase, the medoids are updated by finding a new medoid for each cluster, which minimizes the total distance to other instances in its cluster. In the last phase, the instances are reassigned to their closest medoids.

3.7   Variation of k-Means

3.7.1   *K-Medians*

The k-median algorithm is a variation of the k-means algorithm, which uses the median instead of the mean in order to represent the clusters' centroids. Moreover, the K-median algorithm uses the Manhattan distance as opposed to the Euclidean distances which is used in the K-means algorithm.

The median is calculated for each cluster by independently finding the median of each dimension. Using medians instead of means makes it more robust to outliers compared to K-means. In particular Estivill-Castro (2000) analyzed the performance of the total absolute error criterion. He concluded that while this criterion is superior to SSE regarding robustness, it requires more computational effort.

The k-median algorithm is implemented in *flexclust* package. List 3.9 illustrates this implementation on the Iris dataset. Line 3 defines the dataset for clustering as all attributes in the Iris dataset except the last attribute (column 5). In line 4, clustering is performed on two clusters (k=2). In line 5, the resulting medians of these two clusters are printed. Finally, in line 6, the clustering assignment of all instances is printed.

Listing 3.9: Clustering Iris dataset with K-Medians algorithm

```
1 library("flexclust")
2
3 x=iris[,-5]
4 cl=kcca(x, k=2, family=kccaFamily("kmedians"))
5 print (cl@centers)
6 print (cl@cluster)
```

3.7.2   *K-Modes*

The K-modes algorithm is one of the k-means variations that are designed to deal with nominal (categorical) features [Huang (1998)]. Given a set of instances $\mathbf{X}$ such that each instance is described by n nominal features. The mode of $\mathbf{X}$ is n dimensions vector d not necessarily a member of $\mathbf{X}$ such that the total distance:

$$\mathbf{T}D = \sum_{\mathbf{x} \in \mathbf{X}} \mathbf{d}(\mathbf{x}, \mathbf{d}) \qquad (3.15)$$

where d is a distance measure for categorical attributes as those defined in Section 2.6. Conceptually, the algorithm k-modes works similarly to k-means but instead of calculating clusters' centroids, it finds clusters modes.

The k-modes algorithm is implemented in the *klaR* package. List 3.10 illustrates this implementation. In this example, we use the Arthritis dataset that is provided as part of the *vcd* package. This dataset contains the results of a double-blind clinical trial investigating a new treatment for rheumatoid arthritis. Each row represents one patient who is characterized by the following attributes: ID (patient ID), the treatment (a categorical attribute that can get either "Placebo" or "Treated"), the gender (a categorical attribute that can get either "Female" or "Male"), Age (a numeric attribute for representing the patient's age), Improved (a categorical attribute which indicates the treatment outcome. It can be "None", "Some" or "Marked"). In line 5, the clustering is performed on the arthritis dataset using only the attributes in columns 2,3 and 5 (namely the three categorical attributes: treatment, gender and improved). In this case, we look for two clusters (as indicated by the parameter *modes=2*). In line 6, the resulted modes are printed, and in line 7, the clustering assignment of all instances is printed.

Listing 3.10: Clustering Arthritis dataset with K-modes algorithm

```
1  library("klaR")
2  library("vcd")
3
4  ## run kmodes algorithm on x:
5  cl <- kmodes(Arthritis[,c(2,3,5)], modes=2)
6  print(cl$modes)
7  print(cl$cluster)
```

3.7.3   K-Prototypes

Using the *k*-means algorithm is limited to numeric attributes. The previous section presented the *k-modes* algorithm to address categorical attributes. Haung (1998) presented the *k*-prototypes algorithm, which combines the merit of k-means and k-mode. The algorithm clusters objects with numeric and categorical attributes in a way similar to the *k*-means algorithm. The similarity measure for numeric attributes is the square Euclidean distance; the similarity measure for the categorical attributes is the number of mismatches between the objects and the cluster prototypes. The K-prototypes algorithm is implemented in *R* as the function *pclust*, which is included in the emphclue package.

3.8 The BFR Algorithm

The Bradley-Fayyad-Reina (BFR) algorithm is a scalable clustering method designed to handle very large datasets efficiently. It extends the k-means clustering algorithm by incorporating techniques to manage data that cannot fit into main memory, making it suitable for processing data that are stored on disk or streamed continuously.

The BFR algorithm operates in a two-phase process: the first phase involves summarizing the data using sufficient statistics, and the second phase updates the cluster centroids based on these summaries. This approach allows BFR to handle large datasets incrementally.

3.8.1 *Sufficient statistics*

The BFR algorithm uses three types of sufficient statistics for each cluster:

- N: The number of points assigned to the cluster.
- **SUM**: The vector sum of all points in the cluster.
- **SUMSQ**: The sum of squares of all points in the cluster.

These statistics allow the algorithm to efficiently calculate each cluster's centroid and covariance matrix without storing all the data points.

3.8.2 *Point sets in BFR*

BFR uses three distinct sets to manage data points:

- **Discard set (DS)**: Contains points that are close to their cluster centroids and can be represented by the sufficient statistics. These points are effectively "discarded" after their statistics are computed.
- **Compression set (CS)**: Contains points that are close to each other but not close enough to any existing cluster centroid. These points are temporarily compressed using sufficient statistics until they can be assigned to a cluster or merged with other sets.
- **Retained set (RS)**: Contains points that are outliers or cannot be assigned to any cluster or compression set. These points are retained for future processing.

3.8.3 *Algorithm phases*

The BFR algorithm consists of the following phases:

- **Initialization:** Initialize the algorithm by randomly selecting a small subset of the data to form initial clusters.
- **Summarization:** For each new batch of data, summarize the data points using sufficient statistics.
- **Cluster update:** Update the cluster centroids based on the summarized data. Merge clusters if necessary and discard outliers.

The following subsections present each phase in detail.

3.8.3.1 *Initialization*

1. Select an initial subset of the data that fits into memory.
2. Apply the k-means algorithm to the initial subset to obtain initial cluster centroids.
3. Compute the initial sufficient statistics $(N, \mathbf{SUM}, \mathbf{SUMSQ})$ for each cluster.

3.8.3.2 *Summarization*

For each new batch of data points:

- **Discard set (DS):** Assign each data point to the nearest cluster centroid. Update the sufficient statistics $(N, \mathbf{SUM}, \mathbf{SUMSQ})$ for each cluster.
- **Compression set (CS):** For points that do not fit into any cluster, but are close to each other, form a compression set. Update the sufficient statistics for the compression sets.
- **Retained set (RS):** Points that cannot be assigned to any cluster or compression set are retained for future processing.

3.8.3.3 *Cluster update*

The cluster update phase involves refining the clusters using the summarized information from the discard, compression, and retained sets. This process ensures that the clusters are accurately represented and updated based on the latest data.

1. **Updating clusters from discard set (DS):**

- Compute the new centroids of the clusters using the updated sufficient statistics from the discard set:

$$\boldsymbol{\mu}_i = \frac{\mathbf{SUM}_i}{N_i} \qquad (3.16)$$

- Update the covariance matrices if necessary using the sufficient statistics:

$$\Sigma_i = \frac{\mathbf{SUMSQ}_i}{N_i} - \mu_i \mu_i^T \tag{3.17}$$

2. **Handling compression set (CS):**

- For each compression set, check if it can be merged with an existing cluster. This is done by computing the Mahalanobis distance between the centroid of the compression set and the existing clusters.
- If the compression set is close enough to an existing cluster, merge it by updating the sufficient statistics of the cluster.
- If not, keep the compression set as is or consider creating a new cluster if it consistently appears as a separate group.

3. **Processing retained set (RS):**

- For each point in the retained set, determine if it can now be assigned to a cluster or compression set based on the updated centroids and statistics.
- Points that still cannot be assigned may be kept in the retained set for further batches or treated as outliers.

4. **Cluster merging and splitting:**

- Check for clusters that are too close to each other and merge them if necessary. This helps in reducing redundancy and improving the quality of clustering.
- Optionally, if a cluster shows high variance and might represent multiple clusters, consider splitting it using a method like k-means.

3.8.4 *Applications and use cases*

The BFR algorithm is particularly suited for:

- Clustering very large datasets that cannot fit into main memory.
- Incremental clustering where data arrive in batches or streams.
- Applications requiring efficient computation of cluster statistics for large-scale data.

3.8.5 *Advantages and limitations of BFR algorithm*

The Bradley-Fayyad-Reina (BFR) algorithm is a powerful and scalable method for clustering very large datasets. By leveraging sufficient statistics, the BFR algorithm efficiently summarizes and updates cluster properties, making it suitable for a wide range of applications in data mining and machine learning. Despite its limitations, the BFR algorithm remains a valuable tool for large-scale data analysis.

BFR offers several advantages for data clustering:

- **Scalability**: The BFR algorithm is designed to handle large datasets by summarizing data in memory-efficient ways.
- **Efficiency**: Using sufficient statistics allows for efficient updates and computation of cluster properties.
- **Flexibility**: BFR can process data in batches, making it suitable for both static and dynamic datasets.

However, BFR also has limitations:

- **Initialization sensitivity**: The algorithm's performance can be sensitive to the choice of initial clusters.
- **Merging threshold**: Determining an appropriate threshold for merging clusters can be challenging and may require domain-specific knowledge.
- **Handling outliers**: The method for handling outliers may need to be tailored to the specific application.

3.9 Canopy Clustering

Traditional clustering algorithms like K-Means demand prior knowledge of the number of clusters, which is often unavailable in practical scenarios. Canopy Clustering addresses this by introducing a two-step process that efficiently narrows down potential cluster candidates, thereby facilitating subsequent clustering stages.

Canopy Clustering operates by first placing data points into overlapping subsets, or "canopies", using a computationally cheap distance metric. These canopies provide a rough initial grouping that significantly reduces the number of comparisons required in later, more computationally intensive clustering algorithms. This dual-phase approach, encompassing rough and fine clustering, is highly effective in handling large datasets.

The primary steps of the Canopy Clustering algorithm can be summarized as follows:

(1) Select a random data point and form a canopy by including all points within a loose threshold distance.
(2) Remove the points within a tighter threshold distance from the dataset to avoid redundant canopy formation.
(3) Repeat the process until all points are assigned to at least one canopy.
(4) Perform a secondary, more precise clustering algorithm within each canopy to finalize the cluster assignments.

3.9.1   Detailed description of the algorithm

The algorithm begins by defining two distance thresholds, T_1 and T_2 ($T_1 > T_2$). These thresholds govern the canopy formation process, with T_1 creating broad canopies and T_2 ensuring overlap is minimized. The algorithm iteratively selects data points and forms canopies based on the distance thresholds:

(1) Randomly select a data point, p.
(2) Create a new canopy centered at p.
(3) Add all points within distance T_1 of p to the canopy.
(4) Remove points within distance T_2 of p from the dataset.
(5) Repeat until no points remain that can form new canopies.

Once canopies are formed, a more computationally intensive clustering algorithm, such as K-Means, is applied within each canopy. This step refines the initial rough clustering provided by the canopies.

The pseudo-code of Canopy Clustering is provided in Figure 3.9. It can be broken down into the following steps:

- **Line 0:** Define the algorithm's input parameters, including the dataset D and distance thresholds T_1 and T_2.
- **Line 1:** Initialize an empty list C to store the canopies.
- **Line 2:** Continue the process until all points have been processed.
- **Line 3:** Select a random point p from the dataset.
- **Line 4:** Initialize a new canopy centered around p.
- **Lines 5-12:** For each point q in the dataset, check if it lies within the distance T_1 from p. If so, add q to the current canopy. Additionally, if q lies within distance T_2 from p, remove q from the dataset to prevent redundant canopy formation.
- **Line 13:** Add the newly formed canopy to the list of canopies C.
- **Line 16:** Return the final list of canopies.

Require: Dataset D, distance thresholds T_1 and T_2
Ensure: Set of canopies
  1: Initialize list of canopies $C \leftarrow []$
  2: **while** $D \neq \emptyset$ **do**
  3:   Select a random point $p \in D$
  4:   Initialize a new canopy $canopy \leftarrow \{p\}$
  5:   **for** each point $q \in D$ **do**
  6:     **if** distance$(p, q) \leq T_1$ **then**
  7:       Add q to $canopy$
  8:     **end if**
  9:     **if** distance$(p, q) \leq T_2$ **then**
 10:       Remove q from D
 11:     **end if**
 12:   **end for**
 13:   Add $canopy$ to C
 14: **end while**
 15:
 16: **return** C

Fig. 3.9: Canopy clustering.

3.9.2 *Conclusions, advantages, limitations, and future research*

Canopy Clustering offers a robust solution for efficiently clustering large datasets by leveraging a two-phase approach. It balances computational efficiency with clustering accuracy by forming canopies using a simple distance metric and then refining clusters through a more precise method. This algorithm is particularly useful when dealing with datasets with unknown clusters or where initial rough clustering is required to guide more detailed analyses. The Canopy Clustering algorithm presents several key advantages:

- **Efficiency:** The use of a simple distance metric for initial clustering reduces computational complexity.
- **Scalability:** Suitable for large datasets where traditional clustering algorithms may be impractical.
- **Flexibility:** Can be combined with other clustering algorithms to enhance performance.

Despite its strengths, Canopy Clustering has certain limitations:

- **Choice of thresholds:** The performance depends heavily on the selection of $T1$ and $T2$, which may require domain knowledge or trial-and-error.
- **Initial random selection:** The random selection of initial points can lead to variability in results.
- **Overlap sensitivity:** Overlapping canopies can sometimes complicate the final clustering stage.

Future research directions for Canopy Clustering could focus on:

- **Adaptive thresholds:** Developing methods to dynamically adjust $T1$ and $T2$ based on data characteristics.
- **Robustness improvements:** Enhancing the algorithm's robustness to noise and outliers.
- **Integration with other algorithms:** Exploring the integration of Canopy Clustering with emerging clustering techniques to improve overall performance.
- **Parallelization:** Investigating parallel and distributed implementations to boost efficiency further.

3.10   The k-SVD Algorithm

This section delves into the k-Singular Value Decomposition (k-SVD) algorithm, a powerful method for dictionary learning and sparse representation. We provide an overview of the algorithm, a detailed description including pseudo-code, and instructions on its implementation in R. We conclude with a discussion of its advantages, limitations, and potential areas for future research. The k-SVD algorithm, an extension of the k-means clustering, is a popular method used for dictionary learning and sparse coding. Unlike traditional clustering algorithms, k-SVD is designed to not only group data points but also to learn a dictionary that can sparsely represent these points. This dual functionality makes k-SVD particularly effective in handling high-dimensional data where traditional clustering methods might fall short.

3.10.1   *Detailed Description of the Algorithm*

Below is the pseudo-code for the k-SVD algorithm. The algorithm begins by taking inputs including a data matrix Y, the number of dictionary atoms K, a sparsity level T, and a maximum number of iterations `max_iter`. The outputs are a learned dictionary D and a sparse code matrix X. The process starts by initializing D with random atoms and X with zeros. The main loop iterates up to `max_iter` times. Within this loop, each sample y_i in Y undergoes sparse coding, where the sparse representation x_i is computed by minimizing the reconstruction error under a sparsity constraint. The resulting x_i is stored in X. Then, each atom in the dictionary is updated by first identifying the set of samples that use that atom, computing a residual error matrix excluding the current atom's contribution, and performing Singular Value Decomposition (SVD) on this error matrix. The atom is updated to the first column of the left singular matrix, and its corresponding sparse codes are adjusted accordingly. The atom is then normalized to have a unit norm. Optionally, convergence can be checked by evaluating the change in the dictionary and sparse codes. Finally, the learned dictionary D and sparse codes X are returned as the output.

Require: Data matrix Y, number of dictionary atoms K, sparsity level T, maximum iterations *max_iter*

Ensure: Dictionary D, sparse code matrix X

1: Initialize dictionary D with random atoms
2: Initialize sparse code matrix X with zeros
3: **for** $i = 1$ to *max_iter* **do**
4:     **for** each sample y_i in Y **do**
5:         Solve sparse coding problem: $x_i = \arg\min \|y_i - Dx_i\|_2^2$ subject to $\|x_i\|_0 \leq T$
6:         Store the solution x_i in the corresponding column of X
7:     **end for**
8:     **for** each atom d_k in D **do**
9:         Find the set $\Omega_k = \{i \mid x_{ik} \neq 0\}$
10:         Compute the error matrix $E_k = Y - DX + d_k x_k$
11:         Perform SVD on E_k: $[u, \Sigma, v] = \text{SVD}(E_k)$
12:         Update d_k as the first column of u
13:         Update x_k as the first row of $v\Sigma_{1,1}$
14:         Normalize d_k to unit norm
15:     **end for**
16:     Optional: Check for convergence
17: **end for**
18: **return** D, X

Fig. 3.10: k-SVD Algorithm

3.10.2   Conclusions, advantages, limitations, and future research

In conclusion, the k-SVD algorithm provides a powerful framework for dictionary learning and sparse representation, extending the capabilities of traditional clustering methods. Its dual functionality of simultaneously clustering data points and learning a dictionary makes it particularly effective in high-dimensional data analysis. The k-SVD algorithm presents several key advantages:

- **Dictionary learning:** Unlike traditional clustering algorithms, k-SVD learns a dictionary that can sparsely represent the data points, enabling efficient data compression and feature extraction.
- **Sparse representation:** The algorithm provides a sparse representation of the data, which can be useful in various applications like image processing, signal processing, and natural language processing.
- **Versatility:** k-SVD can be applied to various types of data, including image, audio, and text data, making it a versatile tool for data analysis.

Despite its strengths, k-SVD algorithm has certain limitations:

- **Computational complexity:** The k-SVD algorithm can be computationally expensive, especially for large datasets, due to its iterative nature and the need to perform SVD at each iteration.
- **Sensitivity to initialization:** The algorithm's performance can be sensitive to the initial choice of the dictionary, requiring careful initialization strategies to ensure convergence to a good solution.
- **Parameter selection:** The choice of the number of atoms k and the maximum number of iterations t_{max} can significantly impact the algorithm's performance, requiring careful parameter tuning.

Future research directions could focus on:

- **Efficient implementation:** Developing more efficient implementations of the k-SVD algorithm, leveraging parallel computing and GPU acceleration, can help mitigate its computational complexity.
- **Robust initialization:** Investigating robust initialization strategies for the dictionary can help improve the algorithm's performance and convergence.
- **Theoretical analysis:** Further theoretical analysis of the k-SVD algorithm, including convergence guarantees and bounds on the quality of the learned dictionary, can provide deeper insights into its behavior.

In summary, while the k-SVD algorithm has its limitations, its ability to learn a dictionary and provide sparse representations makes it a valuable tool in various data analysis tasks, with ample opportunities for future research and development.

3.11   Kernel K-Means

Kernel K-Means is a variant of the classical K-Means algorithm that extends it to handle non-linear data by implicitly mapping the data points to a higher-dimensional feature space using a kernel function. This allows K-Means to find clusters in data that may not be linearly separable in the original space. The primary steps of the Kernel K-Means algorithm can be summarized as follows:

(1) Choose a kernel function to map data into a higher-dimensional feature space.
(2) Initialize cluster centroids randomly in the feature space.
(3) Assign each data point to the nearest centroid based on the kernel-induced distance.
(4) Update centroids by computing the mean of assigned points in the feature space.
(5) Repeat the assignment and update steps until convergence.

3.11.1   *Detailed description of the algorithm*

Let $\phi(x)$ represent the mapping of a data point x into the higher-dimensional feature space. The kernel function $K(x, y) = \langle \phi(x), \phi(y) \rangle$ computes the inner product in this feature space. Given a dataset $X = \{x_1, x_2, \ldots, x_n\}$, the objective of Kernel K-Means is to minimize the within-cluster sum of squares in the feature space.

The steps of the Kernel K-Means algorithm can be broken down as follows:

(1) **Initialization:** Randomly initialize cluster centroids $\mu_1, \mu_2, \ldots, \mu_k$ in the feature space.
(2) **Assignment step:** For each data point x_i, assign it to the cluster whose centroid is nearest in terms of the kernel-induced distance:

$$\text{assign}(x_i) = \arg\min_{j} \|\phi(x_i) - \mu_j\|^2$$

(3) **Update step:** Update each centroid μ_j to the mean of all points assigned to cluster j:

$$\mu_j = \frac{1}{|C_j|} \sum_{x_i \in C_j} \phi(x_i)$$

(4) **Convergence check:** Repeat the assignment and update steps until the cluster assignments no longer change.

The pseudo-code for Kernel K-Means Clustering is given in Figure 3.11. It can be broken down into the following detailed steps:

- **Line 1:** Define the input parameters, including the dataset X, number of clusters k, and kernel function K.
- **Line 2:** Initialize the centroids $\mu_1, \mu_2, \ldots, \mu_k$ randomly in the feature space.
- **Line 3:** Begin the iterative process of clustering.
- **Line 4:** Assign each data point to the nearest cluster based on the kernel-induced distance.
- **Line 5:** For each cluster, update the centroid.
- **Line 6:** Compute the new centroid as the mean of all points assigned to the cluster.
- **Line 7:** Repeat until the cluster assignments do not change, indicating convergence.
- **Line 9:** Return the final cluster assignments.

Require: Dataset X, number of clusters k, kernel function K
Ensure: Cluster assignments
1: Initialize centroids $\mu_1, \mu_2, \ldots, \mu_k$ randomly in the feature space
2: **repeat**
3:    Assign each point x_i to the nearest cluster using kernel-induced distance
4:    **for** each cluster j **do**
5:       Update centroid μ_j as the mean of all points assigned to cluster j
6:    **end for**
7: **until** convergence
8:
9: **return** cluster assignments

Fig. 3.11: Kernel K-means clustering.

3.11.2   *Conclusions, advantages, limitations, and future research*

Kernel K-means enhances the traditional K-means algorithm by leveraging kernel methods to handle non-linear separations in the data. This extension allows for more flexible and accurate clustering in complex datasets, making it a powerful tool in the field of unsupervised learning.

The Kernel K-means algorithm presents several key advantages:

- **Non-linear clustering:** It can capture complex, non-linear relationships in the data.
- **Flexibility:** Various kernel functions can be used to adapt to different data structures.
- **Robustness:** Improved clustering performance in scenarios where traditional K-means algorithm fails.

Despite its strengths, Kernel K-means has certain limitations:

- **Computational cost:** Higher computational requirements due to the kernel matrix calculation.
- **Parameter tuning:** Requires careful selection and tuning of the kernel function and its parameters.
- **Scalability:** May struggle with very large datasets due to the kernel matrix size.

Future research directions for Kernel K-means could focus on:

- **Efficient computation:** Developing more efficient algorithms to compute the kernel matrix.
- **Automatic parameter selection:** Creating methods for automatic selection and tuning of kernel parameters.
- **Scalability improvements:** Exploring scalable implementations for handling large datasets.
- **Integration with deep learning:** Investigating the integration of kernel methods with deep learning techniques to enhance clustering performance.

3.12 Mini-Batch K-Means

In the realm of unsupervised learning, clustering algorithms play a pivotal role in the analysis and interpretation of data. While traditional K-means is a cornerstone of clustering techniques, it faces challenges with large datasets due to its computational complexity. Mini-Batch K-means addresses these challenges by introducing a more efficient approach that maintains the integrity of the clustering process while significantly reducing computation time.

Mini-Batch K-means operates by iteratively refining clusters using small, randomly selected subsets (mini-batches) of the data. This approach reduces computational load and improves scalability without sacrificing much accuracy. The primary steps of the Mini-Batch K-means algorithm are:

(1) Initialize cluster centroids randomly.
(2) For each iteration, select a random mini-batch from the dataset.
(3) Assign each point in the mini-batch to the nearest centroid.
(4) Update centroids based on the points assigned in the mini-batch.
(5) Repeat until convergence or for a fixed number of iterations.

3.12.1   *Detailed description of the algorithm*

Let $X = \{x_1, x_2, \ldots, x_n\}$ represent the dataset, where x_i is a data point in $\mathbb{R}^d$. The goal is to partition X into k clusters by minimizing the within-cluster sum of squares. Mini-Batch K-means modifies the traditional K-Means objective function by introducing mini-batches, thereby reducing the amount of data processed in each iteration.

The steps of the Mini-Batch K-means algorithm can be broken down as follows:

(1) **Initialization:** Randomly initialize cluster centroids $\mu_1, \mu_2, \ldots, \mu_k$.
(2) **Mini-batch selection:** Select a random subset (mini-batch) of data points.
(3) **Assignment step:** For each point in the mini-batch, assign it to the nearest centroid:

$$\text{assign}(x_i) = \arg\min_j \|x_i - \mu_j\|^2$$

(4) **Update step:** Update centroids based on the points assigned in the mini-batch:

$$\mu_j = \mu_j + \eta \sum_{x_i \in C_j} (x_i - \mu_j)$$

where η is a learning rate parameter.
(5) **Convergence check:** Repeat the mini-batch selection, assignment, and update steps until convergence or for a fixed number of iterations.

The pseudo-code for Mini-batch K-means clustering is given in Figure 3.12, and it can be broken down into the following detailed steps:

- **Require** Define the input parameters, including the dataset X, number of clusters k, mini-batch size b, learning rate η, and number of iterations T.

- **Line 1:** Initialize the centroids $\mu_1, \mu_2, \ldots, \mu_k$ randomly.
- **Line 2:** Begin the iterative process of clustering.
- **Line 3:** Select a random mini-batch of size b from the dataset.
- **Lines 4-6:** For each point in the mini-batch, assign it to the nearest centroid.
- **Line 7-9:** Update the centroids based on the points assigned in the mini-batch.
- **Line 10:** Repeat the mini-batch selection, assignment, and update steps for T iterations or until convergence.
- **Line 12:** Return the final cluster assignments.

Require: Dataset X, number of clusters k, mini-batch size b, learning rate η, number of iterations T
Ensure: Cluster assignments
1: Initialize centroids $\mu_1, \mu_2, \ldots, \mu_k$ randomly
2: **for** t = 1 to T **do**
3:     Select a random mini-batch of size b from X
4:     **for** each point x_i in the mini-batch **do**
5:         Assign x_i to the nearest centroid
6:     **end for**
7:     **for** each cluster j **do**
8:         Update centroid μ_j
9:     **end for**
10: **end for**
11:
12: **return** cluster assignments

Fig. 3.12: Mini-batch K-means clustering.

3.12.2   *Running the algorithm in R*

List 3.11 illustrates an implementation of the Mini-Batch K-Means algorithm in R. The key components in the R code include:

- **Loading libraries:** The `ClusterR` library is used for clustering, and `ggplot2` is used for visualization.
- **Generating data:** Sample data is generated to demonstrate the algorithm.

- **Running the algorithm:** The `MiniBatchKmeans` function from the `ClusterR` package is called with the sample data, number of clusters, and mini-batch size.
- **Printing results:** The resulting clusters are printed to the console.

Listing 3.11: Using Mini-Batch K-Means for clustering

```
1  library(ClusterR)
2
3  # Generate sample data
4  set.seed(123)
5  n <- 1000
6  data <- data.frame(
7      x = c(rnorm(n, mean = 0, sd = 0.5), rnorm(n, mean = 3, sd
          = 0.5)),
8      y = c(rnorm(n, mean = 0, sd = 0.5), rnorm(n, mean = 3, sd
          = 0.5))
9  )
10
11 # Run Mini-Batch K-Means
12 k <- 2
13 batch_size <- 100
14 clusters <- MiniBatchKmeans(as.matrix(data), clusters = k,
        batch_size = batch_size, num_init = 1, max_iters = 100)
15
16 # Print the results
17 print(clusters$centroids)
```

3.12.3 *Conclusions, advantages, limitations, and future research*

Mini-Batch k-means is a powerful extension of the traditional k-means algorithm that addresses the challenges of clustering large datasets. By processing small, random subsets of data, it achieves significant computational efficiency while maintaining clustering accuracy, making it suitable for large-scale data analysis tasks.

The Mini-Batch K-means algorithm presents several key advantages:

- **Efficiency:** Reduced computational load due to processing small mini-batches instead of the entire dataset.

- **Scalability:** Suitable for large datasets and can handle streaming data.
- **Flexibility:** Can be integrated with other algorithms and used in various applications requiring efficient clustering.

Despite its strengths, Mini-Batch K-means has certain limitations:

- **Approximation:** The algorithm provides an approximation of the true clusters, which may slightly reduce accuracy.
- **Parameter tuning:** Requires careful selection of mini-batch size and other hyperparameters.
- **Initialization sensitivity:** The quality of the clustering can be sensitive to the initial placement of centroids.

Future research directions for Mini-Batch K-means could focus on:

- **Adaptive methods:** Developing adaptive techniques for selecting mini-batch size and learning rate.
- **Enhanced initialization:** Investigating better initialization strategies to improve clustering quality.
- **Integration with deep learning:** Exploring the integration of Mini-Batch K-Means with deep learning frameworks to enhance clustering performance in high-dimensional spaces.
- **Real-time clustering:** Enhancing the algorithm to support real-time clustering for streaming data applications.

3.13 Affinity Propagation

Affinity propagation (AP) is a clustering algorithm that identifies exemplars among data points and forms clusters by assigning each point to its closest exemplar. This algorithm is particularly notable for its ability to handle non-convex clusters. It does not require the number of clusters to be specified beforehand, making it a robust and flexible choice for unsupervised learning.

AP operates by exchanging real-valued messages between data points until a good set of exemplars and corresponding clusters emerges. The messages update in a way that reflects the evidence for one point being the exemplar of another.

3.13.1 *Detailed description of the algorithm*

The pseudocode of the Affinity Propagation algorithm is given in Figure 3.13. The input is a similarity matrix S where $S(i,k)$ indicates how well point i suits to be represented by point k. The algorithm begins by initializing responsibility and availability matrices to zero. Then, the algorithm runs iteratively until convergence. For each data point i and candidate exemplar k, update the responsibility $R(i,k)$. This update reflects the relative suitability of point k as the exemplar for point i. For each candidate exemplar k, update the self-availability $A(k,k)$. Then, for each data point i (except k), update the availability $A(i,k)$. This update ensures that the availability reflects the accumulated evidence for k being an exemplar. Finally, the algorithm outputs the identified exemplars and the corresponding clusters.

1: **Input:** Similarity matrix S
2: Initialize responsibility $R(i,k) = 0$ and availability $A(i,k) = 0$
3: **while** not converged **do**
4:    **for** each data point i **do**
5:       **for** each candidate exemplar k **do**
6:          $R(i,k) \leftarrow S(i,k) - \max\limits_{k' \neq k}\{A(i,k') + S(i,k')\}$
7:       **end for**
8:    **end for**
9:    **for** each candidate exemplar k **do**
10:       $A(k,k) \leftarrow \sum\limits_{i' \neq k} \max\{0, R(i',k)\}$
11:       **for** each data point $i \neq k$ **do**
12:          $A(i,k) \leftarrow \min\{0, R(k,k) + \sum\limits_{i' \notin \{i,k\}} \max\{0, R(i',k)\}\}$
13:       **end for**
14:    **end for**
15: **end while**
16: **Output:** Exemplars and clusters

Fig. 3.13: Affinity propagation.

3.13.2 Running the algorithm in R

To run Affinity Propagation in R, you can use the "apcluster" package. Here is a sample code to demonstrate its usage:

```
Listing 3.12: Affinity Propagation clustering

1  # Install the apcluster package
2  install.packages("apcluster")
3
4  # Load the package
5  library(apcluster)
6
7  # Create a similarity matrix
8  set.seed(123)
9  data <- matrix(rnorm(100), ncol=2)
10 sim_matrix <- negDistMat(data, r=2)
11
12 # Run Affinity Propagation
13 ap_result <- apcluster(sim_matrix)
14
15 # View the results
16 cat("Exemplars:\n")
17 print(ap_result@exemplars)
18 cat("Clusters:\n")
19 print(ap_result@clusters)
```

This script creates a random dataset, computes a similarity matrix, runs Affinity Propagation, and prints the exemplars and clusters.

3.13.3 Conclusions, advantages, limitations, and future research

Affinity propagation offers several advantages:

- It does not require the number of clusters to be pre-determined.
- It handles non-convex clusters well.
- It identifies exemplars, providing interpretable clustering results.

However, it has some limitations:

- It can be computationally intensive for large datasets.
- The choice of similarity function can significantly affect results.
- Sensitivity to hyperparameters, particularly the preference parameter.

Future research directions include:

- Enhancing the efficiency of the algorithm for large datasets.
- Developing methods to automatically tune hyperparameters.
- Extending the algorithm to handle streaming data.

3.14   Fuzzy Clustering

Fuzzy clustering is a type of clustering method that allows data points to belong to multiple clusters with varying degrees of membership. Unlike traditional (hard) clustering methods, where each data point is assigned to exactly one cluster, fuzzy clustering provides a more flexible approach that can capture the inherent ambiguity in the data.

This is particularly useful in scenarios where the boundaries between clusters are not well-defined or when the data contains noise and overlapping clusters. Fuzzy clustering techniques assign a membership value to each data point for each cluster, indicating the degree to which the data point belongs to that cluster.

3.14.1   *Fuzzy C-means (FCM) algorithm*

The fuzzy C-means (FCM) algorithm is one of the most widely used fuzzy clustering methods. It generalizes the k-means algorithm by incorporating fuzzy membership values. The FCM algorithm aims to minimize an objective function representing the weighted sum of squared errors between data points and cluster centers, weighted by the membership values.

3.14.1.1   *Algorithm explanation*

The FCM algorithm iteratively updates the cluster centers and membership values until convergence. The steps are as follows:

(1) **Initialization:** Initialize the membership matrix U randomly, where U_{ij} represents the membership of the i-th data point in the j-th cluster. Ensure that the sum of memberships for each data point is 1, i.e., $\sum_{j=1}^{C} U_{ij} = 1$.

(2) **Cluster center update:** Compute the cluster centers C_j using the following equation:

$$C_j = \frac{\sum_{i=1}^{N}(U_{ij})^m x_i}{\sum_{i=1}^{N}(U_{ij})^m}$$

where m is the fuzziness parameter, N is the number of data points, and C is the number of clusters.

(3) **Membership update:** Update the membership matrix U using the following equation:

$$U_{ij} = \frac{1}{\sum_{k=1}^{C}\left(\frac{\|x_i - C_j\|}{\|x_i - C_k\|}\right)^{\frac{2}{m-1}}}$$

(4) **Convergence check:** Repeat steps 2 and 3 until the changes in the membership values and cluster centers are below a predefined threshold.

The objective function to be minimized is given by:

$$J_m = \sum_{i=1}^{N}\sum_{j=1}^{C}(U_{ij})^m \|x_i - C_j\|^2$$

3.14.1.2 *Parameter selection*

The performance of the FCM algorithm depends on the choice of the fuzziness parameter m and the number of clusters C. Typically, m is chosen to be in the range $[1.5, 3]$. The number of clusters can be determined using methods like the elbow method, silhouette analysis, or cross-validation.

3.14.2 *Implementing FCM in R*

R provides several packages for implementing fuzzy clustering, with the `fclust` package being a popular choice. List 3.13 is an example of implementing the FCM algorithm.

Listing 3.13: Clustering dataset with fuzzy clustering algorithm

```
1  # Install and load the fclust package if not already
       installed
2  if (!require("fclust")) {
3    install.packages("fclust")
4  }
5  library(fclust)
6
7
8  ## McDonald's data
9  data(Mc)
10 names(Mc)
11 ## data normalization by dividing the nutrition facts by the
       Serving Size (column 1)
12 for (j in 2:(ncol(Mc)-1))
13   Mc[,j]=Mc[,j]/Mc[,1]
14
15 ## removing the column Serving Size
16 Mc=Mc[,-1]
17
18 ## fuzzy k-means
19 ## (excluded the factor column Type (last column))
20 clust=Fclust(Mc[,1:(ncol(Mc)-1)],k=6,type="standard",ent=
       FALSE,noise=FALSE,stand=1,distance=FALSE)
21
22 ## fuzzy k-means with polynomial fuzzifier
23 ## (excluded the factor column Type (last column))
24 clust=Fclust(Mc[,1:(ncol(Mc)-1)],k=6,type="polynomial",ent=
       FALSE,noise=FALSE,stand=1,distance=FALSE)
25
26 ## fuzzy k-means with entropy regularization
27 ## (excluded the factor column Type (last column))
28 clust=Fclust(Mc[,1:(ncol(Mc)-1)],k=6,type="standard",ent=
       TRUE,noise=FALSE,stand=1,distance=FALSE)
29
30 ## fuzzy k-means with noise cluster
31 ## (excluded the factor column Type (last column))
32 clust=Fclust(Mc[,1:(ncol(Mc)-1)],k=6,type="standard",ent=
       FALSE,noise=TRUE,stand=1,distance=FALSE)
```

3.14.3 Applications of fuzzy clustering

Fuzzy clustering has a wide range of applications across different fields. Some notable applications include:

3.14.3.1 *Image segmentation*

Fuzzy clustering is used in image segmentation to partition an image into regions with similar properties. The flexibility of fuzzy clustering allows for more accurate segmentation, especially in images with ambiguous boundaries or noise.

3.14.3.2 *Pattern recognition*

In pattern recognition, fuzzy clustering helps identify patterns and structures in the data that are not easily separable. It is used in handwriting recognition, speech recognition, and bioinformatics.

3.14.3.3 *Market segmentation*

Fuzzy clustering is applied in market segmentation to identify groups of customers with similar preferences and behaviors. The fuzzy membership values provide insights into the degree of belonging of each customer to different market segments.

3.14.3.4 *Medical diagnosis*

In medical diagnosis, fuzzy clustering helps classify patients into different categories based on their symptoms and test results. The soft assignments allow for handling the inherent uncertainty in medical data.

3.14.4 *Conclusion*

Fuzzy clustering, with its ability to handle ambiguity and overlapping clusters, provides a powerful tool for data analysis in various domains. The Fuzzy C-means algorithm is a popular and effective method that is easily implemented in R using available packages. Its applications range from image processing to market analysis, demonstrating its versatility and importance in modern data science.

3.15   FLAME Clustering Algorithm

Fuzzy clustering by Local Approximation of MEmberships (FLAME) is a data clustering algorithm designed to effectively handle complex datasets with noise and outliers. It defines clusters in the dense regions of a dataset and approximates the membership of data points to these clusters, making it particularly suitable for real-life applications where clusters may not be well-separated or clearly defined.

FLAME works by first identifying dense regions in the data, which are then used to build a cluster structure. The algorithm uses local information to iteratively update the membership values of data points until convergence.

3.15.1   *Detailed description of the algorithm*

The pseudo-code of FLAME algorithm is given in Figure 3.14. The input consists of data points X and a neighborhood parameter k which determines the number of neighbors considered for density calculation. We begin by computing the k-nearest neighbors for each data point to understand local structures and, calculating the density for each data point based on distances to its neighbors and identifying regions of high and low data point concentration. Then, we identify cluster cores (high-density points) and outliers (low-density points) to form the initial cluster structure and initialize membership values for all points, starting with known cluster cores and outliers. Subsequently, the algorithm iterates, updating membership values using local information until convergence. For each data point x_i, update membership values using a local approximation of the memberships of neighboring points. Once convergence is reached, the algorithm outputs the final membership values and cluster assignments.

1: **Input:** Data points X, neighborhood parameter k
2: Compute the k-nearest neighbors for each data point
3: Calculate the density for each data point based on the distances to its neighbors
4: Identify the cluster cores (high-density points) and outliers (low-density points)
5: Initialize membership values for all points
6: **while** not converged **do**
7:    **for** each data point x_i **do**
8:       Update membership values using a local approximation
9:    **end for**
10: **end while**
11: **Output:** Final membership values and cluster assignments

Fig. 3.14: FLAME clustering.

3.15.2 *Conclusions, advantages, limitations, and future research*

FLAME clustering offers several advantages:

- It effectively handles datasets with noise and outliers.
- It does not require the number of clusters to be specified in advance.
- It provides a soft clustering output, giving membership values reflecting the association degree with clusters.

However, there are limitations to consider:

- It can be computationally intensive for large datasets.
- The performance depends on the choice of the neighborhood parameter k.
- It may struggle with datasets where clusters have very different densities.

Future research directions include:

- Developing more efficient algorithms for large-scale datasets.
- Creating automated methods for parameter selection.
- Extending the algorithm to dynamic and streaming data environments.

3.16 The Gath-Geva Clustering Algorithm

The Gath-Geva (GG) clustering algorithm stands out because it can handle data with varying shapes and densities. Unlike traditional clustering methods, GG does not assume that clusters are spherical or of similar size. Instead, it leverages fuzzy clustering techniques to provide a more flexible and realistic representation of the underlying data structures.

Developed by I. Gath and A.B. Geva in 1989, the GG algorithm is an extension of the Fuzzy C-Means (FCM) algorithm. It incorporates the covariance matrix of each cluster, allowing it to adapt to the cluster's shape and spread. This adaptability makes the GG algorithm particularly useful for applications in pattern recognition and image processing, where data distributions are often irregular.

The GG algorithm combines elements of fuzzy clustering with statistical pattern recognition, resulting in a robust method for partitioning datasets with complex cluster structures. In this chapter, we will delve into the workings of the GG algorithm, provide detailed pseudo-code, and demonstrate its implementation in R.

3.16.1   *Detailed description of the algorithm*

The GG algorithm follows a series of steps to partition data into clusters, considering both the membership degree of each data point to clusters and the covariance structure of the clusters. The algorithm iterates through updates of cluster centroids, membership degrees, and covariance matrices until convergence. Figure 3.15 shows the pseudo-code for the GG algorithm:

Require: $X = \{x_1, x_2, \ldots, x_N\}$ (dataset), C (number of clusters), m (fuzziness parameter), ϵ (convergence threshold)

Ensure: Cluster centroids $\{v_1, v_2, \ldots, v_C\}$, membership matrix U, covariance matrices $\{S_1, S_2, \ldots, S_C\}$

1: Initialize membership matrix U randomly

2: **repeat**

3:    Update cluster centroids v_i:

$$v_i = \frac{\sum_{k=1}^{N} u_{ik}^m x_k}{\sum_{k=1}^{N} u_{ik}^m}$$

4:    Update covariance matrices S_i:

$$S_i = \frac{\sum_{k=1}^{N} u_{ik}^m (x_k - v_i)(x_k - v_i)^T}{\sum_{k=1}^{N} u_{ik}^m}$$

5:    Update membership degrees u_{ik}:

$$u_{ik} = \frac{1}{\sum_{j=1}^{C} \left(\frac{\det(S_i)^{1/2} \exp\left(-\frac{1}{2}(x_k - v_i)^T S_i^{-1}(x_k - v_i)\right)}{\det(S_j)^{1/2} \exp\left(-\frac{1}{2}(x_k - v_j)^T S_j^{-1}(x_k - v_j)\right)} \right)^{\frac{1}{m-1}}}$$

6:    Compute objective function J_m:

$$J_m = \sum_{i=1}^{C} \sum_{k=1}^{N} u_{ik}^m \left[\log(\det(S_i)) + (x_k - v_i)^T S_i^{-1}(x_k - v_i) \right]$$

7: **until** $\|U^{new} - U^{old}\| < \epsilon$

Fig. 3.15: Gath-Geva clustering algorithm.

Now, let's break down each step of the algorithm.

3.16.1.1   Initialization

The algorithm begins by randomly initializing the membership matrix U. This matrix contains the membership degrees of each data point x_k to each cluster i, and its elements u_{ik} satisfy the conditions:

$$0 \le u_{ik} \le 1, \quad \sum_{i=1}^{C} u_{ik} = 1.$$

3.16.1.2   Updating cluster centroids

The cluster centroids v_i are updated based on the membership degrees and the fuzziness parameter m. The new centroid v_i is the weighted average of all data points, with weights given by u_{ik}^{m}.

3.16.1.3   Updating covariance matrices

The covariance matrix S_i for each cluster is updated to reflect the spread and shape of the cluster. It is calculated as the weighted sum of the outer products of the difference between each data point and the cluster centroid.

3.16.1.4   Updating membership degrees

The membership degrees u_{ik} are updated based on the Mahalanobis distance between each data point and the cluster centroids, adjusted by the determinant of the covariance matrix. This step ensures that the algorithm accounts for the shape and orientation of each cluster.

3.16.1.5   Objective function

The objective function J_m measures the overall clustering quality by combining the log determinant of the covariance matrices and the Mahalanobis distances. The algorithm aims to minimize this objective function through iterative updates.

3.16.1.6   Convergence

The algorithm iterates through these steps until the change in the membership matrix U is less than a predefined threshold ϵ.

3.16.2   *Running the algorithm in R*

To run the Gath-Geva algorithm in R, you can use the "ppclust" package. The script below illustrates this idea. The script aims to cluster Iris dataset. It runs the Gath-Geva clustering algorithm and prints the membership values of the top 5 samples.

Listing 3.14: Using Gath-Geva algorithm for clustering

```r
1  # Load the package
2  library(ppclust)
3
4  # Load dataset iris
5  data(iris)
6  x <- iris[,-5]
7
8  # Initialize the prototype matrix using Inofrep algorithm
9  v <- inaparc::inofrep(x, k=3)$v
10 # Initialize the memberships degrees matrix
11 u <- inaparc::imembrand(nrow(x), k=3)$u
12
13 # Run Gath & Geva with the initial prototypes and
       memberships
14 gg.res <- gg(x, centers=v, memberships=u, m=2)
15
16 # Show the fuzzy memberships degrees for the top 5 objects
17 head(gg.res$u, 5)
```

3.16.3   *Conclusions, advantages, limitations, and future research*

The Gath-Geva clustering algorithm presents a powerful approach to clustering data with complex structures. Its main advantages include:

- Adaptability to non-spherical and varied cluster shapes.
- Consideration of cluster covariance, providing more realistic partitions.
- Robustness in handling noise and outliers.

However, the algorithm also has some limitations:

- Increased computational complexity compared to simpler methods like K-means.

- Sensitivity to initialization, which may affect convergence and final results.
- Requirement for a predefined number of clusters, similar to many other clustering methods.

Future research could focus on several areas to enhance the GG algorithm:

- Developing more efficient initialization techniques to improve convergence.
- Extending the algorithm to handle large-scale datasets through parallel computing or other optimization methods.
- Integrating with deep learning approaches to enhance clustering performance in high-dimensional spaces.

In conclusion, the Gath-Geva algorithm is a versatile and robust tool for clustering complex data structures, and ongoing research continues to expand its applicability and efficiency.

3.17 Gustafson-Kessel Clustering

The Gustafson-Kessel (GK) clustering algorithm is an extension of the Fuzzy C-Means (FCM) clustering method, which identifies clusters with different geometrical shapes. This is achieved by modeling each cluster with its own covariance matrix, thereby enabling the clusters to take various forms such as ellipsoidal shapes. This flexibility makes the GK algorithm particularly useful for datasets where the underlying cluster structures are not spherical.

The GK algorithm belongs to the family of fuzzy clustering methods. In fuzzy clustering, each data point can belong to multiple clusters with varying degrees of membership, which are quantified by membership values. This approach provides a more nuanced understanding of the data structure compared to hard clustering methods, where each data point is assigned to exactly one cluster.

3.17.1   *Detailed description of the algorithm*

The core idea of the GK algorithm is to minimize an objective function that incorporates the covariance matrices of the clusters. Let $X = \{x_1, x_2, \ldots, x_N\}$ be a set of N data points in d-dimensional space. The algorithm aims to partition this set into C clusters by minimizing the following objective function:

$$ J = \sum_{i=1}^{C} \sum_{k=1}^{N} u_{ik}^{m} D_{ik}^{2} $$

where u_{ik} is the membership degree of the k-th data point in the i-th cluster, m is the fuzziness parameter, and D_{ik} is the Mahalanobis distance between the k-th data point and the i-th cluster center.

The algorithm iteratively updates the cluster centers, covariance matrices, and membership degrees until convergence. The pseudocode is presented below and it consists of the following steps:

(1) **Initialization:** The algorithm starts by initializing the cluster centers v_i and the membership matrix U. This can be done randomly or using some heuristic.

(2) **Cluster center update:** The cluster center v_i is updated as the weighted average of all data points, with weights given by the membership degrees raised to the power of m.

(3) **Covariance matrix update:** The covariance matrix F_i is updated to reflect the spread of the data points around the cluster center.

(4) **Distance calculation:** The Mahalanobis distance D_{ik} is computed for each data point with respect to each cluster.

(5) **Membership degree update:** The membership degrees u_{ik} are updated based on the distances to all cluster centers, ensuring that the membership degrees sum to 1 for each data point.

(6) **Convergence check:** The algorithm repeats the update steps until convergence, typically defined as when changes in the membership matrix fall below a certain threshold.

3.17.2   *Running the algorithm in R*

You can use the "fclust" package to run the Gustafson-Kessel algorithm in R. The script given in Figure 3.16. illustrates this idea. The script generates random data, runs the Gustafson-Kessel clustering algorithm, and prints the identified cluster centers, membership values, and covariance matrices.

1: Initialize the cluster centers v_i and the membership matrix U.

2: **repeat**

3:    **for** each cluster i **do**

4:       Update the cluster center v_i:

$$v_i = \frac{\sum_{k=1}^{N} u_{ik}^m x_k}{\sum_{k=1}^{N} u_{ik}^m}$$

5:       Compute the covariance matrix F_i:

$$F_i = \frac{\sum_{k=1}^{N} u_{ik}^m (x_k - v_i)(x_k - v_i)^T}{\sum_{k=1}^{N} u_{ik}^m}$$

6:       Compute the inverse covariance matrix F_i^{-1}.

7:    **end for**

8:    **for** each data point k **do**

9:       **for** each cluster i **do**

10:          Update the distance D_{ik}:

$$D_{ik}^2 = (x_k - v_i)^T F_i^{-1} (x_k - v_i)$$

11:          Update the membership degree u_{ik}:

$$u_{ik} = \left(\sum_{j=1}^{C} \left(\frac{D_{ik}}{D_{jk}} \right)^{\frac{2}{m-1}} \right)^{-1}$$

12:       **end for**

13:    **end for**

14: **until** convergence

Fig. 3.16: Gustafson-Kessel clustering algorithm.

Listing 3.15: Using Gustafson-Kessel algorithm for clustering

```
1  # Load the package
2  library(fclust)
3
4  # Generate sample data
5  set.seed(123)
6  data <- matrix(rnorm(200), ncol=2)
7
8  # Run Gustafson-Kessel clustering
9  gk_result <- Fclust(data, k=3, type="gk")
10
11 # View the results
12 print(gk_result)
```

3.17.3 *Conclusions, advantages, limitations, and future research*

The Gustafson-Kessel clustering algorithm offers several advantages over traditional clustering methods:

(1) **Flexibility:** By allowing clusters to take different shapes, the GK algorithm can model complex data structures more accurately.

(2) **Fuzzy membership:** The use of fuzzy membership degrees provides a richer representation of data, capturing uncertainty and overlap between clusters.

However, the GK algorithm also has some limitations:

(1) **Computational complexity:** The need to compute and invert covariance matrices increases the computational cost, especially for high-dimensional data.

(2) **Sensitivity to initialization:** Like many clustering algorithms, the GK method is sensitive to the initial choice of cluster centers and membership degrees.

Future research directions for the GK algorithm include:

(1) **Scalability:** Developing scalable versions of the algorithm that can handle large datasets efficiently.

(2) **Robustness:** Enhancing the robustness of the algorithm to noisy data and outliers.

(3) **Automated initialization:** Improving methods for automatically determining the initial parameters to reduce sensitivity to initialization.

In conclusion, the Gustafson-Kessel clustering algorithm is a powerful tool for identifying complex cluster structures in data. Its ability to adapt the shape of clusters and provide fuzzy memberships makes it a valuable addition to the data scientist's toolkit. Continued research and development will further enhance its applicability and performance in various domains.

Chapter 4

Hierarchical Methods

4.1 Introduction

Hierarchical methods aim to build a hierarchy of clusters by recursively partitioning the objects in either a bottom-up or top-down fashion. These methods can be sub-divided as follows:

- Agglomerative hierarchical clustering — Each object starts as a separate cluster. These clusters are then gradually combined into larger groups until the target cluster configuration is achieved.
- Divisive hierarchical clustering — Initially, all objects are grouped into a single cluster. This cluster is then divided into smaller clusters, and these, in turn, are further divided into even smaller ones. This division process is repeated until the desired cluster structure is achieved.

The result of the hierarchical methods is a tree structure called dendrogram. The dendrogram represents the nested grouping of objects (in an agglomerative method) or the nested partitioning (in a divisive method). In addition, the dendrogram shows the similarity levels at which groupings change. A certain clustering of the data can be obtained by cutting the dendrogram at the desired similarity level. Figure 4.8 shows a dendrogram for 12 flowers that were randomly selected from the Iris dataset. At height=0, all twelve flowers are singleton clusters. At height=0.1, flowers 141 and 78 are joined together. Then, at height=0.141, flowers 105 and 138 are joined together. The process continues until all points are merged at height=2.061 into one large cluster.

Clusters are either merged or split based on a specific similarity measure in an attempt to optimize some goal function (like minimizing the sum of squared differences). The methods of hierarchical clustering can be further categorized by how this similarity metric is calculated [Jain *et al.* (1999)]:

- **Single-link clustering** (also called the connectedness, the minimum method or the nearest neighbor method) — methods that determine the distance between two clusters by considering the shortest distance between any member of one cluster and any member of the other cluster. When the data consists of similarities, the similarity between a pair of clusters is defined as the greatest similarity between any member of one cluster and any member of the other cluster [Sneath and Sokal (1972)].
- **Complete-link clustering** (also called the diameter, the maximum method or the furthest neighbor method) — methods that determine the distance between two clusters as the longest distance from any member of one cluster to any member of the other cluster [King (1967)].
- **Average-link clustering** (also called minimum variance method) — methods that use the average distance from any member of one cluster to any member of the other cluster. Such clustering algorithms may be found in [Ward Jr (1963)] and [Murtagh (1983)].

The disadvantages of single-link clustering and the average-link clustering can be summarized as follows [Guha *et al.* (1998a)]:

- Single-link clustering has a drawback known as the "chaining effect": A few points that form a bridge between two clusters cause the single-link clustering to unify these two clusters into one.
- Average-link clustering may cause elongated clusters to split and portions of neighboring elongated clusters to merge.

Complete-link clustering methods typically produce more compact clusters and more meaningful hierarchies compared to single-link clustering methods. However, single-link methods are more versatile in their application. Generally, hierarchical methods are characterized with the following strengths:

- Versatility — The single-link methods, for example, maintain good performance on data sets containing non-isotropic clusters, including well-separated, chain-like and concentric clusters.
- Multiple partitions — As we will illustrate later, many hierarchical methods create multiple nested partitions, which allow different users

to choose different partitions, according to the desired similarity level. The hierarchical partition is presented using the dendrogram.

The main disadvantages of the hierarchical methods are:

- Inability to scale well — The time complexity of hierarchical algorithms is at least $O(m^2)$ (where m is the total number of instances), which is non-linear with the number of objects. Clustering a large number of objects using a hierarchical method is also characterized by huge I/O costs.
- Hierarchical methods can never undo what was done previously. Namely, there is no back-tracking capability.

4.2 Agglomerative Methods

Agglomerative methods work in a bottom-up manner. They begin by forming a separate cluster for each point. In the first step, the two closest entities (according to some similarity measure) are merged, resulting in $M - 1$ clusters, one of which consists of two entities while the other clusters still contain a single entity. The procedure continues by merging the entities or clusters that are closest to each other, until all the entities are merged into one cluster (the topmost level of the hierarchy), or a termination condition is met. The basic pseudo-code of the agglomerative algorithm is presented in Figure 4.1.

Due to the fact that in each iteration, exactly two clusters are merged and that in the beginning, there are m clusters, each of which contains a single instance, thus an agglomerative method needs at most m iterations.

Require: X (instance set)
Ensure: clusters
1: Compute the distance matrix between the points.
2: Set each point as a cluster
3: **repeat**
4:    Merge the two closest clusters.
5:    Recalculate distance matrix among clusters
6: **until** only a single cluster remains

Fig. 4.1: The basic pseudo-code of agglomerative algorithm.

Agglomerative methods are characterized according to the different distance measures that are used to find the closest clusters. Note that the distance measures presented in Chapter 2 cannot be used as-is, because they are defined as a distance between single entities. Here, on the other hand, we need to measure the distance between clusters, where each cluster contains many instances. It is useful to differentiate between two common types of cluster distance measures: graph measures and geometric measures. In the case of graph measures, each cluster is represented as a subgraph and the distance between two clusters is calculated as a function of the distance among the individual entities. In geometric measures, a cluster is represented by a single point (such as the centroid), and the distance between any two clusters is calculated based on the distance among their representing points.

4.2.1 *Graph measures*

In order to illustrate the definitions of various graph measures, we will consider the following case: The dataset contains 12 points obtained from the Iris dataset using only two dimensions (Sepal Length and Sepal Width). Table 4.1 specifies their coordinates. Currently, the points are clustered into 5 clusters as demonstrated in Figure 4.2. Note that each of the points 3 and 12 constitutes its own cluster (cluster C_5 and C_4 respectively).

Table 4.1: Iris dataset.

Point	Sepal Length	Sepal Width
1	7.2	3.2
2	4.9	3.6
3	6.2	2.2
4	6.7	3.1
5	6.7	3.3
6	5.5	2.6
7	5.0	3.4
8	5.6	3.0
9	5.8	2.7
10	4.6	3.6
11	5.7	3.0
12	4.9	3.0

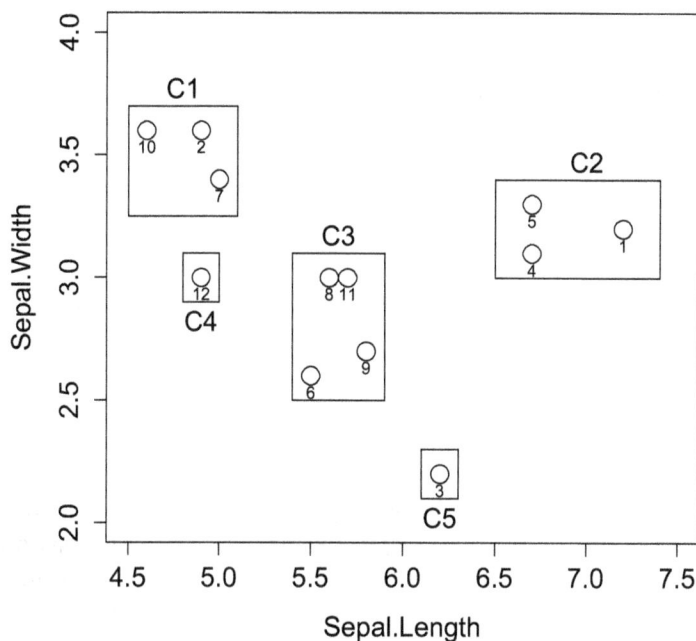

Fig. 4.2: Visualization of 12 points in Iris dataset clustered into 5 clusters.

4.2.1.1 *Single-link*

According to the single-link measure, the dissimilarity between two clusters is defined as the minimum distance among all possible pairs of points that are made up of one data point from each cluster. Specifically, the single-link distance between two nonempty, non-overlapping clusters C_i and C_j is defined as:

$$D_{Single-Link}(C_i, C_j) = \min_{\mathbf{x} \in C_i; \mathbf{y} \in C_j} d(\mathbf{x}, \mathbf{y}) \tag{4.1}$$

where $d(\mathbf{x}, \mathbf{y})$ is the distance between two entities or points. It can be any distance measure presented in Chapter 2, such as the Euclidean distance. Based on the above equation, it is obvious to imagine why the single-link measure is also known by other names: the minimum distance, and the nearest neighbor measure.

In order to find which clusters should be merged in Figure 4.2, we can go over all pairs of clusters and calculate the single link distance. For example the distance between cluster C_1 and C_2 is:

$$D(C_1, C_2) = min(d(\mathbf{x_10}, \mathbf{x_5}), d(\mathbf{x_10}, \mathbf{x_4}), d(\mathbf{x_10}, \mathbf{x_1}), d(\mathbf{x_2}, \mathbf{x_5}),$$
$$d(\mathbf{x_2}, \mathbf{x_4}), d(\mathbf{x_2}, \mathbf{x_1}), d(\mathbf{x_7}, \mathbf{x_5}), d(\mathbf{x_7}, \mathbf{x_4}), d(\mathbf{x_7}, \mathbf{x_1}))$$
$$= min(0.51, 0.582, 0.642, 0.453, 0.524, 0.585, 0.362, 0.409, 0.541)$$
$$= d(x_7, x_5) = 0.362$$

By going over all pairs of clusters, we can find that clusters C_3 and C_4 should be merged as they are the closest pair (the single link distance is only about 0.134).

Obviously, the need to recalculate in each iteration from scratch the distances among all clusters is tedious. To avoid that, it is useful to maintain a dissimilarity matrix that is updated after each iteration. We begin by calculating the distance among all pairs of points. In each iteration, as we merge the two closest two clusters, we update the matrix based on the merged clusters. Specifically, in the case of the single-link distance measure, we merge the corresponding rows and columns of the two clusters by taking the minimum of the two merged values. The current distance matrix for Figure 4.2 is presented in Table 4.2. According to this matrix, the shortest distance (not including zeroes that represent self-distance) is 0.134 (the distance between cluster C_3 and C_4). Thus, we merge these two clusters as presented in Figure 4.3.

Table 4.2: Distance matrix for the Iris dataset.

	C1	C2	C3	C4	C5
C1	0	0.362	0.258	0.162	0.362
C2	0.362	0	0.228	0.381	0.417
C3	0.258	0.228	0	0.134	0.255
C4	0.162	0.381	0.134	0	0.535
C5	0.362	0.417	0.255	0.535	0

In order to update the distance matrix, we merge the columns of clusters 3 and 4 into a single column by taking the minimum for each position in the vectors. Specifically, the columns of clusters 3 and 4 are: $[0.258, 0.228, 0, 0.134, 0.255]$ and $[0.162, 0.381, 0.134, 0, 0.535]$. By merging these two vectors using the minimum, we get: $[0.162, 0.228, 0, 0, 0.255]$. Then, we merge the corresponding rows of clusters 3 and 4 and get the new distance matrix presented in Table 4.3.

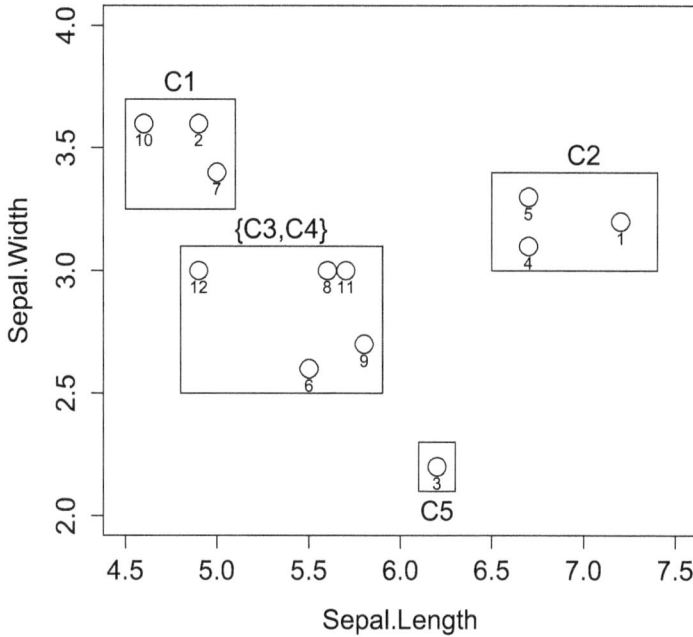

Fig. 4.3: Visualization of 12 points in Iris dataset after merging clusters 3 and 4 using the single-link method.

Table 4.3: Distance matrix for the Iris dataset.

	C1	C2	C3,C4	C5
C1	0	0.362	0.162	0.362
C1	0.362	0	0.228	0.417
C3,C4	0.162	0.228	0	0.228
C5	0.362	0.417	0.255	0

4.2.1.2 *Complete link*

Similarly to the single-link measure, the complete link measure also determines the distance between two clusters by considering the distance between all possible pairs of data points that are made up of one data point from each group.

However, instead of taking the closest pair as the between clusters distance, it takes the farthest pair of points. Thus, the distance measure between two nonempty, non-overlapping clusters C_i and C_j is defined as:

$$D_{\text{Complete Link}}(C_i, C_j) = \max_{\mathbf{x}\in C_i;;\mathbf{y}\in C_j} d(\mathbf{x}, \mathbf{y}) \tag{4.2}$$

where $d(\mathbf{x}, \mathbf{y})$ is the distance between two entities or points. If we use the complete link measure, the distance matrix for Figure 4.2 becomes the matrix presented in Table 4.4. For example, the distance between clusters 1 and 2 is calculated as:

$$D(C_1, C_2) = max\big(d(\mathbf{x_1}0, \mathbf{x_5}), d(\mathbf{x_1}0, \mathbf{x_4}), d(\mathbf{x_1}0, \mathbf{x_1}), d(\mathbf{x_2}, \mathbf{x_5}),$$

$$d(\mathbf{x_2}, \mathbf{x_4}), d(\mathbf{x_2}, \mathbf{x_1}), d(\mathbf{x_7}, \mathbf{x_5}), d(\mathbf{x_7}, \mathbf{x_4}), d(\mathbf{x_7}, \mathbf{x_1})\big)$$

$$= max(0.51, 0.582, 0.642, 0.453, 0.524, 0.585, 0.362, 0.409, 0.541)$$

$$= d(x_7, x_5) = 0.362$$

Table 4.4: Distance matrix for the Iris dataset.

	C1	C2	C3	C4	C5
C1	0	0.642	0.552	0.271	0.807
C2	0.642	0	0.541	0.513	0.549
C3	0.552	0.541	0	0.28	0.401
C4	0.271	0.513	0.28	0	0.535
C5	0.807	0.549	0.401	0.535	0

Given the distance matrix, we merge the closest pair of clusters, i.e., cluster 1 and cluster 4 which have the smallest between-cluster distance of 0.271. Pay attention to the fact that, in this case, the complete link measure has led us to a different decision than we previously made using the single-link measure. Figure 4.4 illustrates the obtained clustering using the complete link method.

In order to update the distance matrix, we merge the columns of clusters 1 and 4 into a single column by taking the maximum for each position in the vectors. Specifically, the columns of clusters 1 and 4 are: $[0, 0.642, 0.552, 0.271, 0.807]$ and $[0.271, 0.513, 0.28, 0, 0.535]$. By merging these two vectors using the maximum function, we get: $[0.271, 0.642, 0.552, 0.271, 0.807]$. Then we similarly merge the corresponding rows of clusters 1 and 4 using the maximum function again. Finally, we set the values on the diagonal to zero. Table 4.5 presents the new distance matrix.

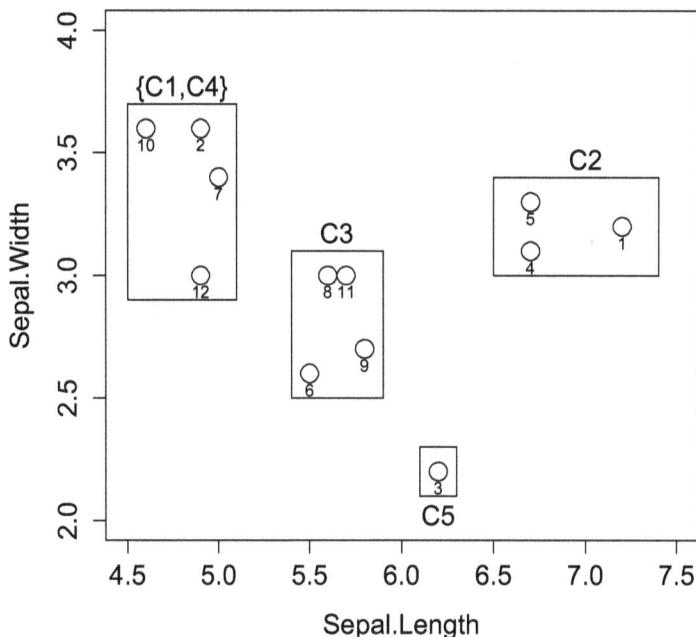

Fig. 4.4: Visualization of 12 points in Iris dataset after merging clusters 3 and 4 using complete link method.

Table 4.5: Distance matrix for the Iris dataset after merging clusters 1 and 4 assuming complete link.

	C1,C4	C2	C3	C5
C1,C4	0.271	0.642	0.552	0.807
C2	0.642	0	0.541	0.549
C3	0.552	0.541	0	0.401
C5	0.807	0.549	0.401	0

4.2.1.3 *Group average*

According to the group average measure, the distance between two clusters is defined as the average of the distances between all pairs of points, one point taken from each cluster.

$$D_{\text{Group Average}}(C_i, C_j) = \frac{1}{|C_i| \cdot |C_j|} \sum_{\mathbf{x} \in C_i; \mathbf{y} \in C_j} d(\mathbf{x}, \mathbf{y}) \qquad (4.3)$$

where C_i and C_j are two nonempty, non-overlapping clusters and $d(\mathbf{x}, \mathbf{y})$ is the distance between two points.

4.2.1.4 McQuitty

In McQuitty's linkage method, the distance of the new cluster to any other cluster is calculated as the average of the distances of the merged clusters to any other cluster. For example, if clusters 1 and 2 are merged into a new cluster, then the distance from the new cluster to cluster 3 is the average of the distances from cluster 1 to cluster 4 and cluster 2 to 3. Specifically, given three nonempty, non-overlapping clusters C_i, C_j and C_k:

$$D(C_i \cup C_j, C_k,) = \frac{1}{2} \times D(C_i, C_k) + \frac{1}{2} \times D(C_j, C_k)$$

McQuitty's linkage method is also known by other names, such as the weighted pair group method using arithmetic average or the weighted group average method.

4.2.2 Geometric measures

In geometric measures, a cluster is represented by a single point (such as the centroid), and the distance between any two clusters is calculated based on the distance among their representing points. After merging the closest pair of clusters, a new representative point is derived directly from the two clusters.

4.2.2.1 Centroid

The centroid linkage method, which is also known as the name unweighted pair group method using centroids, uses the distance between the centroids of the clusters to measure the distance between two clusters. Specifically the distance between any two non-overlapping clusters C_i and C_j, is defined as:

$$\|\mu(C_i) - \mu(C_j)\|^2$$

where $\mu(C)$ denotes the center of cluster C, that is: $\mu(C) = \frac{1}{|C|} \sum_{\mathbf{x} \in C} \mathbf{x}$.

The distance can also be represented using the distance between the individual points as follows:

$$D_{\text{Centroid}}(C_i, C_j)$$

$$= \frac{1}{|C_i| \cdot |C_j|} \sum_{\mathbf{x} \in C_i;; \mathbf{y} \in C_j} d(\mathbf{x}, \mathbf{y}) - \frac{1}{2|C_i|^2} \sum_{\mathbf{x}, \mathbf{y} \in C_i} d(\mathbf{x}, \mathbf{y})$$

$$- \frac{1}{2|C_j|^2} \sum_{\mathbf{x}, \mathbf{y} \in C_j} d(\mathbf{x}, \mathbf{y}) \tag{4.4}$$

When merging the closest pair of clusters, the new merged cluster is represented by a new centroid derived as follows:

$$\frac{|C_i|\mu(C_i) + |C_j|\mu(C_j)}{|C_i| + |C_j|} \tag{4.5}$$

4.2.2.2 *Median*

The median method is also known by the name weighted pair group method using centroids. With the median linkage method, the distance between two clusters is the median distance between an object in one cluster and an object in the other cluster. This can be used to avoid the effect of outliers that exist in the centroid method.

After merging the closest pair of clusters, a new representing point is derived according to the following formula:

$$\frac{\mu(C_i) + \mu(C_j)}{2} \tag{4.6}$$

4.2.2.3 *Ward method*

The goal of Ward's linkage method is to minimize the within-cluster sum of squares. Thus, the distance between two clusters is the sum of squared deviations from points to centroids. Specifically, the distance between any two non-overlapping clusters is defined as: $\frac{|C_i||C_j|}{|C_i|+|C_j|}\|\mu(C_i) - \mu(C_j)\|^2$ where $\mu(C)$ denotes the center of cluster C. After merging the closest pair of clusters, a new represented point of the new cluster should be calculated. In the Ward method, this can be derived directly from that of the two clusters as follows:

$$\frac{|C_i|\mu(C_i) + |C_j|\mu(C_j)}{|C_i| + |C_j|} \tag{4.7}$$

The main advantage of Ward's method is that it tends to produce balanced clusters with similar numbers of points in each cluster. On the other hand, the main drawback of the Ward method is that it is sensitive to outliers.

4.2.3 *Running the basic hierarchical clustering algorithm in R*

In this section, we illustrate how to use the basic hierarchical clustering algorithm that comes with the *stats* package of R to cluster a sample of the Iris dataset. Line 1 of Code 4.1 projects the original dataset into a 2D dataset using only the first two columns.

In line 2 the distance matrix between any pair of flowers is computed using the Euclidean distance function by setting *method= "euclidean"* (which is also the default measure in this case). Other distance measures supported are *"maximum"*, *"manhattan"*, *"canberra"*, *"binary"* or *"minkowski"* (see Chapter 2 for additional details). In line 3, the hierarchical clustering is performed using the group average method. The supported methods are: *"single"* for single-link method; *"complete"* for complete link method; *"average"* for group average method; *"mcquitty"*; *"median"*; *"centroid"* and finally *"ward.D"*, *"ward.D2"* for two different implementation of the Ward method.

The algorithm begins with assigning each flower to its own cluster. Then at each iteration, the algorithm joins the most similar clusters to a single cluster. The iterations continue till we are left with a single cluster, which consists of all flowers in the dataset. In line 4, the cluster tree, or dendrogram, is plotted. It illustrates the hierarchical arrangement of the clusters produced by the clustering. The leaves of the dendrogram (the lowest nodes in the graph) represent the individual flowers. The remaining nodes represent the clusters to which the flowers are assigned. The root of the dendrogram (the node located at the top of the graph) represents the universal cluster that consists of all flowers in the dataset. The height of each node in the graph refers to the distance between its two children.

Listing 4.1: Using basic hierarchical clustering to cluster Iris dataset

```
1 iris2dsample  <- iris[sample(nrow(iris),12),1:2]
2 DistX=dist(iris2dsample,method="euclidean")
3 cl<-hclust(DistX,method="ave")
4 plot(cl)
```

The *fastcluster* package provides another implementation of the agglomerative method. It has the same interface as the original *hclust* function, but it is much faster. Code 4.2 illustrates this implementation.

Listing 4.2: Using fast hclust implementation to cluster Iris dataset

```
1 library(fastcluster)
2 iris2dsample <- iris[sample(nrow(iris),12),1:2]
3 DistX=dist(iris2dsample,method="euclidean")
4 cl<-hclust(DistX,method="ave")
5 plot(cl)
```

4.2.4 ROCK algorithm

ROCK (RObust Clustering using linKs) is an agglomerative hierarchical clustering algorithm that merges clusters by using links [Guha *et al.* (1999)]. ROCK is particularly useful for datasets with Boolean and categorical attributes.

Let $sim()$ be a similarity function between two points that takes a value in $[0, 1]$. Two points $\mathbf{x_i}$ and $\mathbf{x_j}$ are considered to be neighbors if their similarity value exceeds a certain threshold value θ, i.e. $sim(\mathbf{x_i}, \mathbf{x_j}) > \theta$. Let us define $link(\mathbf{x_i}, \mathbf{x_j})$ to be the number of common neighbors between $\mathbf{x_i}$ and $\mathbf{x_j}$. The assumption is that two points with many common neighbors should be assigned to the same cluster. Thus, given two clusters, C_i and C_j their goodness measure is defined as:

$$g(C_i, C_j) = \frac{\sum_{\mathbf{x} \in C_i;; \mathbf{y} \in C_j} link(\mathbf{x}, \mathbf{y})}{(|C_i| + |C_j|)^{1+2f(\theta)} - |C_i|^{1+2f(\theta)} - |C_j|^{1+2f(\theta)}} \qquad (4.8)$$

where $|C_i|^{1+2f(\theta)}$ is the expected number of links between pairs of points in C_i. The function $f(\theta)$ is a positive function such that every point $\mathbf{x}$ in C_i has approximately $|C_i|^{f(\theta)}$ neighbors in C_i. Assuming that points outside C_i has probably a small number of links to the points in C_i, each point in cluster C_i has $|C_i|^{2f(\theta)}$ links one link for each pair of its neighbors. Thus as there are $|C_i|$ points in C_i we conclude that the expected number of links between pairs of points in C_i is $|C_i|^{1+2f(\theta)}$.

Obviously, finding the right function is not an easy task. However, it has been shown that even an inaccurate but reasonable estimate for $f(\theta)$ can work fairly well in practice [Guha *et al.* (1999)]. The inventors of the ROCK algorithm show that the following function can work well in market basket transactions dataset:

$$f(\theta) = \frac{1-\theta}{1+\theta} \qquad (4.9)$$

Figure 4.5 shows the pseudo-code of the ROCK algorithm. The algorithm begins by calculating the links among all pairs of points. The procedure for calculating the links is provided in Figure 4.6. The main ROCK algorithm maintains for each cluster C_i a local heap $q[C_i]$ that contains every cluster C_j such that $g(C_i, C_j) > 0$. The clusters in the local heap are ordered in decreasing order of their goodness measure. The algorithm also maintains a global heap Q that contains all clusters sorted according to their best goodness measure (i.e., the cluster with the greatest goodness measure is located at the top of the heap).

Require: X (instance set), n (number of clusters)
Ensure: clusters
  1: Compute the links function between all points.
  2: For each point $\mathbf{x}$ calculates a local heap $q[\mathbf{x}]$
  3: Build global heap Q
  4: **repeat**
  5:    Get from Q the cluster v with the maximum goodness measure
  6:    Get from $q[\mathbf{v}]$ the cluster u with largest goodness measure
  7:    Join cluster u and v
  8:    Update heaps
  9: **until** n clusters remain or all clusters have zero links

Fig. 4.5: The pseudo-code of the ROCK algorithm.

Require: X (instance set)
Ensure: Links
  1: Compute the list of neighbors for all points in X
  2: Let $link[\mathbf{x_i}, \mathbf{x_j}] = 0$ for $\forall \mathbf{x_i}, \mathbf{x_j} \in X$
  3: **for** $i = 1$ **to** $|X-$ **do**
  4:    Set N to the list of neighbors of $\mathbf{x_i}$
  5:    **for** $j = 1$ **to** $|N|$ - 1 **do**
  6:      **for** $l = j + 1$ **to** $|N|$ **do**
  7:      $link[N[j], N[l]] = link[N[j], N[l]] + 1$
  8:      **end for**
  9:    **end for**
10: **end for**

Fig. 4.6: The pseudo-code for computing the links function between all points.

The *cba* package provides an implementation of the ROCK algorithm. List 4.3 illustrates how it is being used to cluster the Iris dataset. In line 4 we define the distance function to be used by the algorithm. In line 5, the clustering is performed by setting the requested number of clusters to $n=2$ and the neighboring threshold to *theta=0.1*. In line 6 the clustering model is printed and in line 7 the points are plotted where the points symbol reflect their cluster assignment.

Listing 4.3: Using ROCK algorithm to cluster Iris dataset

```
1 library("cba")
2 set.seed(2)
3 iris2dsample <- iris[sample(nrow(iris),100),1:2]
4 gdist <- function(x, y=NULL) 1-exp(-dist(x, y)^2)
5 cl <- rockCluster(iris2dsample, n=2, theta=0.1,fun=gdist,
      funArgs=NULL)
6 print(cl)
7 plot(iris2dsample, pch = as.numeric(cl$cl))
```

4.2.5 AGNES

AGNES (AGglomerative NESting) is a simple hierarchical agglomerative clustering algorithm that follows the pseudo-code presented in Figure 4.1. At first, each entity forms a cluster by itself. At each stage, the two closest clusters are combined to form one larger cluster. Clusters are merged until all the entities are assigned to one big cluster.

The AGNES implementation is provided in R by the *cluster* package. In addition to the standard output, this implementation provides an agglomerative coefficient value, which measures the amount of clustering structure found. Moreover it provides a banner graphical display for the results.

The *agnes* function supports the following clustering methods: *"single"* for the single-link method; *"complete"* for the complete link method, and *"average"* for the group average method. In addition, it supports the flexible Lance-Williams formula, which generalizes all hierarchical agglomerative methods to a single method with four parameters. Specifically, the distance between a new cluster (formed by merging two clusters) and all other clusters is calculated as:

$$D(C_k, C_i \cup C_j) = \alpha_i D(C_k, C_i) + \alpha_j D(C_k, C_j) + \beta D(C_i, C_j) + \gamma |D(C_k, C_i) - D(C_k, C_j)|$$

where $d(\mathbf{C_i}, \mathbf{C_j})$ is the distance between two clusters or entities. By choosing the right values for the parameters α_i, α_j, β, and γ one can obtain various inter-cluster distances used by the hierarchical clustering algorithms. For example, the single-link method can be obtained by setting the parameters as follows: $\alpha_i = 0.5, \alpha_j = 0.5$, $\beta = 0$ and $\gamma = -.5$. Similarly, the complete link method corresponds to the following setting $\alpha_i = 0.5$, $\alpha_j = 0.5$, $\beta = 0$ and $\gamma = .5$. Furthermore, the weighted group average can be obtained by the following setting: $\alpha_i = 0.5$, $\alpha_j = 0.5$, $\beta = 0$ and $\gamma = 0$.

In R, the four coefficients are specified by the vector *par.method* as a parameter of the *agnes* function. The list in Figure 4.4 demonstrates the usage of *agnes* function for clustering the Iris dataset. The output of this script is two graphs. The first graph presented in Figure 4.7 is a banner in which the heights of the dendrogram are represented on a horizontal axis. In addition, it presents the agglomerative coefficient, which measures the clustering structure of the dataset as reflected by the average width of the banner plot. Because the agglomerative coefficient increases with the number of points, it cannot be used to compare datasets of different sizes. Figure 4.8 presents the dendrogram resulting from the clustering. In order to get the exact heights in which entities are merged, you can print the clustering object by using the command *print(cl)*. The *agnes* object can be

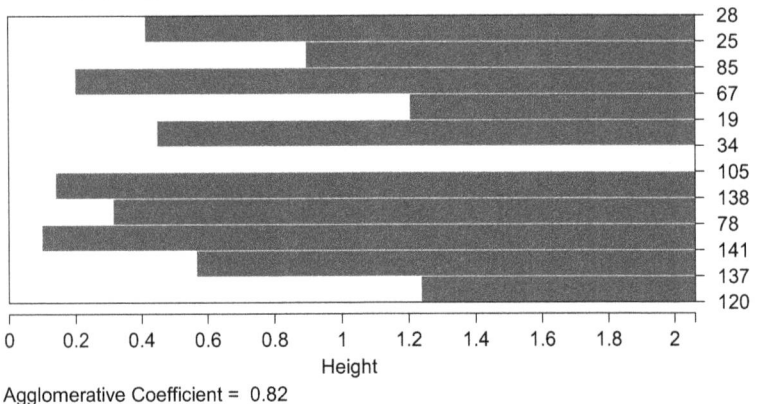

Fig. 4.7: The banner plot obtained for the AGNES clustering.

Dendrogram of agnes(x = DistX, method = "flexible", par.method = c(0.5, 0.5, 0, 0.5))

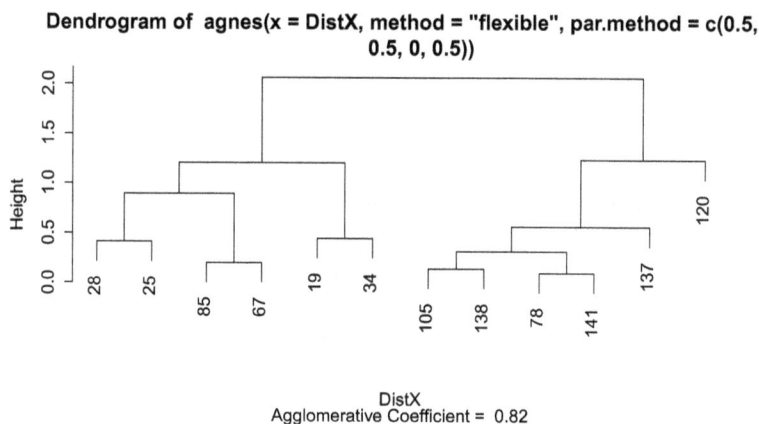

Fig. 4.8: The Dendrogram obtained for the AGNES clustering.

converted to an object of class *hclust* by using the *as.hclust()* function. The conversation to "hclust" can be useful for performing standard analysis.

Listing 4.4: Using AGNES algorithm to cluster Iris dataset

```
1 library(cluster)
2 set.seed(2)
3 iris2dsample <- iris[sample(nrow(iris),12),1:2]
4 DistX=dist(iris2dsample,method="euclidean")
5 cl<-agnes(DistX,method = "flexible", par.method = c
    (0.5,0.5,0,0.5))
6 plot(cl)
```

4.3 Divisive Methods

A divisive hierarchical clustering method uses a top-down strategy. Initially, all of them form an initial root cluster. In each iteration, the method splits the existing cluster into several smaller sub-clusters according to some splitting criterion, such as the maximum Euclidean distance between the nearest neighboring entities in the cluster. This recursive partitioning process continues until, eventually, every cluster consists of a single entity or the entities within a cluster are sufficiently homogeneous.

Partitioning large clusters into sub-clusters can be challenging. If a cluster has n entities, then there are $2^{(n-1)} - 1$ different ways to split it into two sub-cluster.[1] Thus, for large values of n it is intractable to go over all combinations. Therefore, a heuristics method should be employed.

4.3.1   *DIANA*

DIANA (DIvisive ANAlysis) is a popular divisive method [Kaufman and Rousseeuw (2009)] implemented in R. It begins by assigning all entities into one cluster. At each iteration, the cluster with the largest diameter is divided into two sub-clusters. The distance between the farthest points in the cluster determines the diameter of a cluster. More specifically:

$$Diam(C) = \max_{\forall \mathbf{x}, \mathbf{y} \in C} d(\mathbf{x}, \mathbf{y}) \qquad (4.10)$$

After selecting the cluster to be divided, the algorithm searches for an entity in this cluster that is farthest from all other entities in this cluster. More specifically, we look for point y^* which satisfies the following definition:

$$\mathbf{y}^* = \underset{\mathbf{y} \in C}{\operatorname{argmax}} \sum_{\mathbf{x} \in C, \mathbf{x} \neq \mathbf{y}} D(\mathbf{x}, \mathbf{y}) \qquad (4.11)$$

The selected entity $\mathbf{y}^*$ is used to form the so-called splinter sub-cluster A, the remaining entities are assigned to the majority sub-cluster B. Then, for each entity in B, we compute the average distance with the remaining entities in B, and compare it to its average distance with the entities in A:

$$DistDiff(\mathbf{y}) = \frac{1}{|B| - 1} \sum_{\mathbf{x} \in B, \mathbf{x} \neq \mathbf{y}} D(\mathbf{x}, \mathbf{y}) - \frac{1}{|A|} \sum_{\mathbf{z} \in A} D(\mathbf{z}, \mathbf{y}) \qquad (4.12)$$

The point $\mathbf{y}$ which obtains a positive and the largest value for the function $DistDiff()$ is moved from B to A. Because it disagrees more with the points in B than with the points in A. After updating the sub-clusters A and B, the process repeats itself by recalculating the $DistDiff()$ for the remaining points in B. The next point is selected to move from B to A and so forth. This process continues whenever there are points with strictly

---

[1]Note that each entity can be in one of two clusters A and B. There are 2^n assignment combinations, but this number should be divided by two because, for each sub-cluster assignment, there is a complementary assignment that partially represents the same partition (the sub-clusters A and B are switched). In addition, the case in which all entities are assigned to the same cluster is not a valid one.

```
1 library(cluster)
2 set.seed(2)
3 iris2dsample <- iris[sample(nrow(iris),12),1:2]
4 DistX=dist(iris2dsample,method="euclidean")
5 cl<-diana(DistX)
6 plot(cl)
```

Dendrogram of diana(x = DistX)

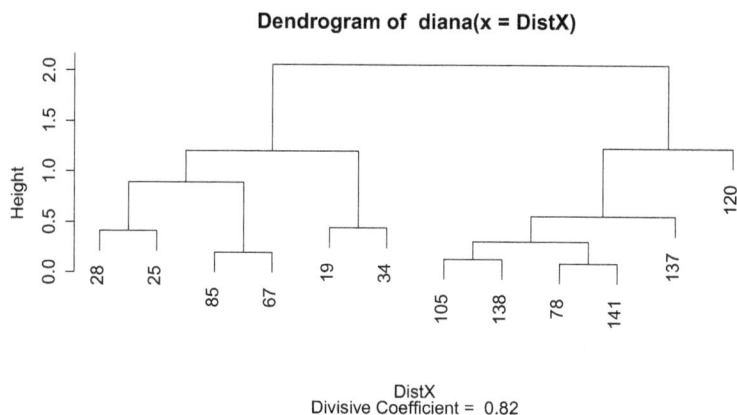

DistX
Divisive Coefficient = 0.82

Fig. 4.9: The Dendrogram obtained for the DIANA clustering.

positive $DistDiff()$ values. If the largest value is negative, the partition of C into A and B is stopped. After we have completed the first divisive step, we run the entire procedure again, i.e., searching for the largest cluster among the current clusters and dividing it into two sub-clusters and so forth.

The DIANA function is provided in R by the *cluster* package. The code 4.5 illustrates its usage for clustering the Iris dataset. Figure 4.9 presents the dendrogram resulting from this clustering. It is interesting to compare it with the dendrogram obtained by the AGNES method (Figure 4.8). In this case, both have obtained the same dendrogram. But, obviously, this will not always be the case.

4.3.2 COBWEB

COBWEB is a hierarchical clustering that incrementally incorporates entities into a tree structure [Fisher (1987)]. Each node in the tree is a probabilistic representation of the entities of which it consists. The algorithm begins with an empty root node, and it incrementally adds entities by traversing the tree top-down and considering the following operations:

Assigning the new entity to an existing node Creating a new node and assigning the new entity to this new node Merging two nodes into a single node and placing the new entity in the resulting hierarchy; Dividing a node into several nodes and placing the new entity in the resulting hierarchy.

COBWEB uses the category utility heuristic measure to compare the intra-cluster similarity and inter-cluster dissimilarity of entities and, based on that, select the appropriate operation to perform on the new entity. In particular, category utility estimates the quality of a given clustering $C_1, \ldots, C_k$ as the improvement in probability estimate of the entities get their correct values given the clustering structure over the probability estimate for getting those values with no clustering [Corter and Gluck (1992)]. Assuming that entities are described in terms of nominal attribute-value pairs (i.e. $A_i = V_{i,j}$), then formally, the category utility is the improvement in probability summed across all clusters (l), attributes (i), and values (j) as follows:

$$CU(C_1, \ldots, C_k)$$

$$= \frac{1}{k} \sum_{l=1}^{k} p(C_l) \left[\sum_{i=1}^{n} \sum_{j} p\left(A_i = V_{i,j} \,|\, C_l\right)^2 - \sum_{i=1}^{n} \sum_{j} p\left(A_i = V_{i,j}\right)^2 \right]$$

$$(4.13)$$

where the term $p(C_l)$ is the prior probability of an entity belonging to the cluster C_l, the term $p\left(A_i = V_{i,j}\right)$ denotes the probability that the i-th feature A_i takes the j-th value $V_{(}i,j)$ and the term $p\left(A_i = V_{i,j} \,|\, C_l\right)$ represents the conditional probability of the same event given the clustering assignments. The term $\sum_{i=1}^{n} \sum_{j} p\left(A_i = V_{i,j} \,|\, C_l\right)^2$ is the expected number of attribute values that can be correctly guessed for an arbitrary entity that belongs to cluster C_l. The denominator, k, which represents the number of clusters, is required to enable comparison of clusterings of different sizes.

While there is no native implementation of COBWEB in CRAN (The Comprehensive R Archive Network), it is still possible to run COBWEB in R using a special interface to Weka included in the *RWeka* package.

Recall that Weka is installed with only the most popular learning algorithms. In order to install other algorithms, one should use the package manager to select which additional component is to be installed. This is done in R using *WPM* function.

Code 4.6 illustrates the process. In the first, the *RWeka* package is loaded. In the second, line the *COBWEB* algorithm is added to Weka. In line 3, we call *Cobweb* algorithm by setting the category utility threshold by which it prunes nodes to 0.002. The complete available list of parameters that can be set, can be obtained by executing the command *WOW("weka/clusterers/Cobweb")*. Note that we do not use the fifth feature, which holds the species class during the clustering. Finally in line 4, we print the results and in line 5 we compare actual flowers species with their cluster assignments obtained via *predict()* function. As can be seen in the results table, Cobweb grouped all flowers of the same species into the same cluster.

Listing 4.6: Using COBWEB algorithm to cluster Iris dataset

```
1 library ("RWeka")
2 WPM("install-package", "Cobweb")
3 cl2 <- Cobweb(iris[, -5], ("-C", 0.002))
4 print(cl2)
5 table(predict(cl2),iris$Species)
```

4.4 Hybrid Hierarchical Clustering

In general, divisive methods are excellent at identifying large clusters but not small ones, while agglomerative methods are excellent at identifying small clusters but not large ones. Hybrid hierarchical clustering methods combine the strengths of agglomerative and divisive methods [Chipman and Tibshirani (2006)].

Chipman and Tibshirani (2006) introduce the notion of a mutual cluster as a group of points that are sufficiently close to each other and distant from all other points. The entities contained in a mutual cluster should never be separated. Formally, a mutual cluster is a subset S of entities, such that:

$$d(\mathbf{x}, \mathbf{y})_{\mathbf{x} \in S, \mathbf{y} \notin S} > diameter(S) \equiv \max_{\mathbf{w} \in S, \mathbf{z} \in S} d(\mathbf{w}, \mathbf{z})$$

where $d()$ is a distance function, $\mathbf{x}$ is an object that belongs to S, and $\mathbf{y}$ is an object that does not belong to S. The above definition ensures that the largest distance between objects in S is smaller than the smallest distance from an object in S to any object not in S.

Assuming Euclidean distance, points 1, 4 and 5 in Figure 4.2 form a mutual cluster. The largest distance between these points is $d(\mathbf{x_1}, \mathbf{x_4}) = d(\mathbf{x_1}, \mathbf{x_5}) = \sqrt{0.5^2 + 0.2^2}$. All distances between $\{1, 4, 5\}$ and $\{2, 3, 6, 7, 8, 9, 10, 11, 12\}$ are greater than the diameter, making $\{1, 4, 5\}$ a mutual cluster. On the other hand, points $\{2, 7, 10\}$ does not form a mutual cluster, because $d(\mathbf{x_7}, \mathbf{x_{10}}) > d(\mathbf{x_7}, \mathbf{x_{12}})$.

The mutual cluster has the following property: it is not broken by an agglomerative clustering method with any of the following methods: single linkage, complete linkage or average linkage. Thus, mutual clusters can be easily found by using the agglomerative clustering method. Moreover, the concept of mutual clusters can be used to create a hybrid method by performing the following steps: Compute the mutual clusters using an agglomerative clustering method. Temporarily replace mutual clusters with their centroids and perform a top-down clustering. In this way all mutual clusters are intact. Partition each mutual cluster by performing a top-down clustering within each mutual cluster.

The *hybridHclust* package uses the notion of mutual cluster to construct a clustering. This package depends on the *cluster* package as it employs top-down and bottom-up methods. The *hybridHclust* package is demonstrated in List 4.7. In line 4, we find the mutual clusters. The dendrogram based on these mutual clusters is then plotted as illustrated in Figure 4.10. In line 5, we print the mutual clusters that were found and their content. As can be seen in the output presented in list 4.8, there are four mutual clusters. The first mutual cluster consists of points 7 and 12. The second mutual cluster consists of points 3 and 9, and so forth. In line 8 we perform the hybrid clustering and in line 10 the resulted dendrogram is displayed as illustrated in Figure 4.11. In line 12 we repeat clustering, this time using pre-calculated mutual clusters.

The *mutualCluster* function has the following arguments:

x – Data matrix to be clustered; *distances* – Distances between objects are usually clustered. Usually produced by *dist* function. Note that either x or *distances* should be provided; *method* – Method used to join clusters. Must be one of single, complete or average. Because all three methods provide the same mutual clusters, this option does not determine the mutual clusters returned but it only affects the plotting; *plot* – Flag indicating if the dendrogram for bottom-up clustering should be presented.

Listing 4.7: Using hybrid hierarchical clustering algorithm to cluster Iris dataset

```
1 library(hybridHclust)
2 set.seed(2)
3 iris2dsample <- iris[sample(nrow(iris),12),1:2]
4 mc1 <- mutualCluster(iris2dsample , plot=TRUE)
5 print(mc1) # print the mutual clusters
6
7 # Perform Hybrid Hierarchical Clustering
8 hyb1 <- hybridHclust(iris2dsample)
9 # Plot the Hybrid Clustering Dendogram
10 plot(hyb1)
11
12 # Perform Hybrid Hierarchical Clustering using precalculated
      mutual clusters
13 hyb2 <- hybridHclust(iris2dsample,mc1)
14 # Plot the Hybrid Clustering Dendogram
15 plot(hyb2)
```

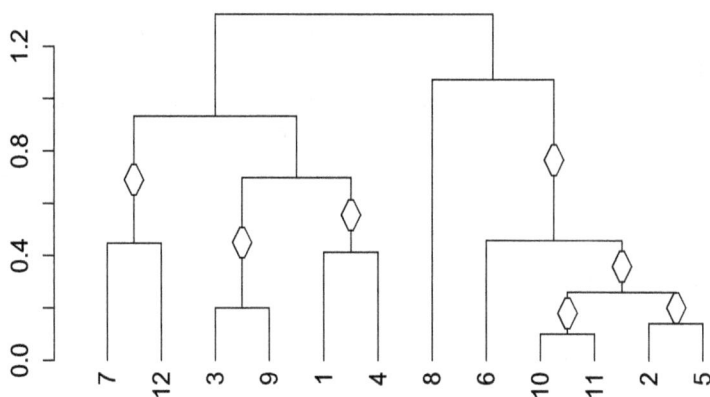

Fig. 4.10: The dendrogram based on mutual clusters.

Listing 4.8: Output of hybrid hierarchical clustering

```
1 1 : 7 12
2 2 : 3 9
3 3 : 1 4
4 4 : 2 5 6 10 11
```

Cluster Dendrogram

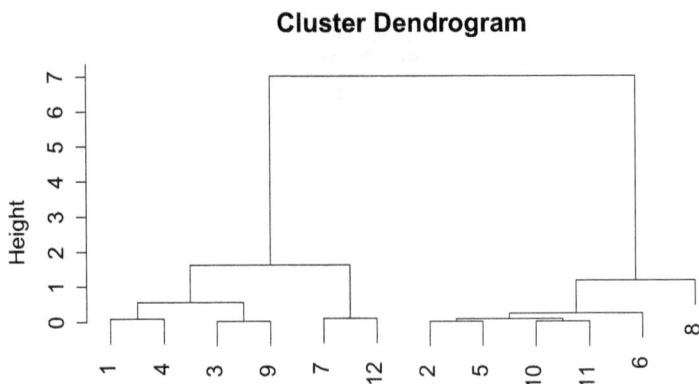

Fig. 4.11: The final dendrogram resulted from the hierarchical clustering of the Iris dataset.

4.5 Supporting Packages

In addition to the implementations of the hierarchical clustering algorithms presented in the previous sections, R has several supporting packages. In this section, we will review and demonstrate two important packages: The *dynamicTreeCut* package contains methods for the detection of clusters in hierarchical clustering dendrograms. The *pvclust* package contains methods for assessing the uncertainty in hierarchical cluster analysis.

4.5.1 *Detection of clusters in hierarchical clustering dendrograms*

The *cutreeDynamic* function performs adaptive branch pruning of hierarchical clustering dendrograms. It gets a hierarchical clustering dendrogram such as one returned by *hclust*. Optionally it also gets a distance matrix that was used for creating the dendorgram. The *cutreeDynamic* function returns a vector of numerical labels giving assignment of objects to clusters. The points that are assigned to the largest cluster have the label 1, the next largest cluster 2 and so forth. Unassigned objects have the label 0.

List 4.9 demonstrates the usage of *cutreeDynamic* function. In line 4, we create the distance matrix. In line 5, the hierarchical clustering is created. The *cutreeDynamic* function is used for cutting the tree. The *deepSplit* argument provides a rough control over sensitivity to cluster splitting. The higher the value is, the smaller clusters will be created. We could set other

Listing 4.9: Using *cutreeDynamic* function to cut a dendorgram

```
1 library(dynamicTreeCut)
2 library(sparcl) # needed for providing colored dendrogram
3 iris2d <- iris[,1:2]
4 DistX=dist(iris2d,method="euclidean")
5 hc <- hclust(DistX, method="centroid")
6 labels = cutreeDynamic(hc, distM =  as.matrix(DistX) ,
    deepSplit = 2)
7
8 # coloring the leaves of a dendrogram
9 ColorDendrogram(hc, y = labels, labels = names(labels))
```

arguments to guide the tree-cutting. In particular, the *cutHeight* argument specifies the cut height and the *minSize* argument specifies the minimum number of elements in a cluster. Finally in line 7 we use the obtained labels to plot a colorful dendorgram using the *ColorDendrogram* function provided in the *sparcl* package. Note that the points' colors are determined according to the assignments obtained by the *cutreeDynamic* function.

4.5.2 *Assessing the uncertainty in hierarchical cluster analysis*

The *pvclust* package provides an implementation of multiscale bootstrap resampling for assessing the uncertainty in hierarchical cluster analysis [Shimodaira *et al.* (2004)]. It calculates p-values for each cluster obtained by the hierarchical clustering. The most important arguments in *pvclust* function are the following: data – data to be clustered. method.hclust – the agglomerative method to be used for the hierarchical clustering. We can select one of the following implementation in the *hclust* package: "average", "ward", "single", "complete", "mcquitty", "median" or "centroid".

method.dist – the measure to be used to measure the dissimilarities between objects. It can be any method that is allowed in *dist* function (i.e. *"euclidean"*, *"maximum"*, *"manhattan"* etc.). In addition, it can be one of the following "correlation" which takes uncentered sample correlation, "uncentered" which takes uncentered sample correlation or "abscor" takes the absolute value of sample correlation.

nboot – the number of bootstrap replications. The default is 1000.

In line 5, the *pvclust* function is executed. Note that we provide the transpose of the data. In line 7, we print the result of multiscale bootstrap resampling. Line 9 plots a dendrogram with the p-values which are calculated by the multiscale bootstrap resampling. Figure 4.12 displays this plot.

```
1 library(pvclust)
2 set.seed(2)
3 iris2dsample <- iris[sample(nrow(iris),12),1:4]
4 # multiscale bootstrap resampling
5 hc <- pvclust(t(iris2dsample) , method.hclust="single",
      method.dist="euclidean",nboot=100)
6 # print the result of multiscale bootstrap resampling
7 print(hc)
8 # plot dendrogram with p-values
9 plot(hc,cex.pv=1)
```

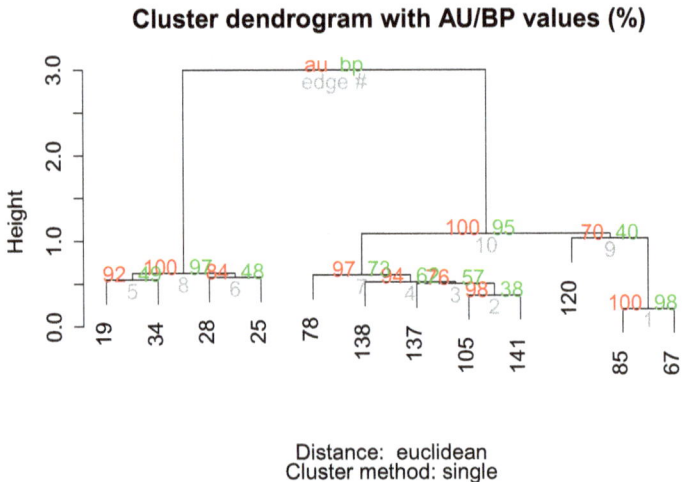

Cluster dendrogram with AU/BP values (%)

Distance: euclidean
Cluster method: single

Fig. 4.12: A dendrogram resulted with p-values. The values in red represent "approximately unbiased" p-values. The values in green represent "bootstrap probability" values.

The values in red represent "approximately unbiased" p-values (or AU for short). The values in green represent "bootstrap probability" values (or BP for short). Clusters (edges) with high AU values (say 95%) are strongly supported by the data.

4.6 The BIRCH Algorithm

The BIRCH (Balanced Iterative Reducing and Clustering using Hierarchies) algorithm is a scalable and efficient method for clustering large datasets. Developed by Tian Zhang, Raghu Ramakrishnan, and Miron Livny in 1996, BIRCH is designed to handle "very large datasets" by incrementally and dynamically clustering incoming, multi-dimensional metric data points in a single scan of the dataset. This capability makes it particularly suited for data mining applications.

BIRCH uses a hierarchical data structure called the CF (Clustering Feature) tree for clustering. The CF tree is a height-balanced tree that stores the clustering features for a hierarchical clustering algorithm. The BIRCH algorithm proceeds in two main phases:

(1) Scans the data and builds an initial in-memory CF tree.
(2) Applies a selected clustering algorithm to the leaf entries of the CF tree to cluster the data.

4.6.1 *Detailed description of the algorithm*

The BIRCH algorithm can be described in terms of its key steps and components. Figure 4.13 shows the pseudo-code for the BIRCH algorithm. First, we initialize an empty CF tree T with parameters B and L, where the parameter B represents the branching factor of the tree (i.e., the maximum number of children a node can have), and the parameter L represents the maximum number of entries in a leaf node.

For each data point x_i: we iterate over each data point in the dataset. Insert x_i into the appropriate leaf of the CF tree T. To determine the best leaf for insertion based on the minimum increase in the CF distance metric. Moreover, we check if the insertion of x_i causes the leaf to exceed its capacity. If the leaf exceeds its capacity, split the leaf and adjust the tree structure accordingly. Once the CF tree is constructed, apply a global

1: Initialize an empty CF tree T with parameters B and L
2: **for** each data point x_i **do**
3:     Insert x_i into the appropriate leaf of the CF tree T
4:     **if** necessary **then**
5:         Split the leaf and adjust the CF tree
6:     **end if**
7: **end for**
8: Apply a global clustering algorithm to the leaf entries of the CF tree T

Fig. 4.13: BIRCH algorithm.

clustering algorithm such as k-means to the leaf entries to produce the final clusters.

4.6.2   *CF tree structure*

A CF tree is a height-balanced tree composed of non-leaf nodes and leaf nodes. Each node contains a list of entries, and each entry is a clustering feature (CF) that summarizes information about a cluster. The CF is defined as a triple (N, LS, SS), where N is the number of data points in the cluster, LS is the linear sum of the data points, and SS is the squared sum of the data points.

The CF tree construction involves inserting data points into the tree, updating the CFs, and splitting nodes if necessary. The following sections explain these processes in detail.

4.6.3   *Insertion into CF tree*

Inserting a data point into the CF tree involves finding the appropriate leaf node and updating the CFs. If the insertion causes a leaf node to exceed its capacity, the node is split, and the tree is adjusted to maintain its balance.

4.6.4   *Node splitting*

When a leaf node exceeds its capacity, it is split into two nodes. The parent node is updated to reflect the split, and if the parent node also exceeds its capacity, it is split recursively. This process ensures that the CF tree remains balanced.

4.6.5   *Conclusions, advantages, limitations, and future research*

The BIRCH algorithm offers several advantages and limitations, which are discussed below. Additionally, potential areas for future research are highlighted.

The BIRCH algorithm has several notable advantages:

(1) Scalability: It is highly efficient for large datasets due to its single-scan and multi-phase clustering approach.
(2) Incremental Clustering: It can incrementally update clusters as new data points are added.

(3) Hierarchical Clustering: The CF tree structure provides a hierarchical clustering framework, allowing for multi-level analysis.

Despite its advantages, the BIRCH algorithm also has some limitations:

(1) Sensitivity to Order of Data: The algorithm's results can be sensitive to the order in which data points are processed.
(2) Parameter Selection: Choosing appropriate values for the branching factor and threshold can be challenging and may require domain knowledge.
(3) Not Suitable for All Data Types: The algorithm is primarily designed for metric data and may not perform well on categorical data.

Future research directions for the BIRCH algorithm include:

(1) Adaptive Parameter Tuning: Developing methods for automatically tuning the algorithm's parameters to improve clustering performance.
(2) Extension to Categorical Data: Extending the algorithm to handle categorical or mixed-type data.
(3) Integration with Other Algorithms: Combining BIRCH with other clustering algorithms to enhance its robustness and accuracy.

In conclusion, the BIRCH algorithm is a powerful tool for clustering large datasets, offering a balance between efficiency and effectiveness. With ongoing research and development, its applicability and performance can be further enhanced.

4.7 SLINK Algorithm: A Dive into Hierarchical Clustering

Hierarchical clustering stands as a cornerstone in the realm of data clustering, offering a rich perspective on data structure through nested partitions. Among its techniques, the Single-Linkage Clustering (SLINK) algorithm presents a compelling approach, emphasizing proximity-based merging to construct a hierarchical dendrogram. In this section, we embark on an odyssey through the intricacies of SLINK, from its conceptual underpinnings to practical implementations in R. At the heart of SLINK lies a simple yet powerful principle: the merging of clusters based on the smallest pairwise distance between points. This principle echoes the notion of "nearest neighbors", where clusters converge gradually, forming a hierarchical structure akin to the branches of a dendrogram.

4.7.1   *Detailed description of the algorithm*

The pseudocode given in Figure 4.14 presents the outline of the SLINK algorithm. First, we initialize arrays L, S, and M to manage cluster assignments, nearest neighbors, and inter-cluster distances, respectively. Then, we sort the pairwise distances in ascending order to facilitate efficient merging. Subsequently, we iterate over the sorted distances, merging clusters based on proximity. Finally, we update the nearest neighbor and minimum distance arrays accordingly.

Initialize an array L of length n with each element representing a singleton cluster.
Initialize an array S of length n with each element representing the nearest neighbor of each cluster.
Initialize an array M of length n to store the minimum inter-cluster distances.
Sort the distances in D in ascending order.
for $i = 1$ **to** $n - 1$ **do**
   Merge clusters L_i and L_j where L_j is the nearest neighbor of L_i.
   for $k = 1$ **to** n **do**
      if $M_k > d(L_i, L_k)$ **then**
         Update M_k to $d(L_i, L_k)$.
         Update S_k to L_i.
      end if
   end for
end for

Fig. 4.14: SLINK algorithm.

4.7.2   *Conclusions*

In conclusion, the SLINK algorithm offers a robust framework for hierarchical clustering, characterized by its simplicity and efficiency. Its advantages lie in its ability to handle large datasets and capture intricate data structures. However, like any algorithm, SLINK is not devoid of limitations, particularly its sensitivity to noise and outlier data. Future research endeavors may explore enhancements to mitigate these shortcomings and extend the algorithm's applicability to diverse domains.

4.8 The CLINK (Complete-Linkage Clustering)

Among the family of agglomerative hierarchical clustering methods, complete-linkage clustering stands out for its ability to produce compact, spherical clusters that are resilient to noise and outliers. However, the classical implementation suffers from a computational complexity of $O(n^3)$, rendering it impractical for large-scale applications. This is where the CLINK algorithm shines, providing an optimally efficient solution with a reduced complexity of $O(n^2)$.

Developed by D. Defays in the 1970s, CLINK is an ingenious algorithm that builds upon the principles of complete-linkage clustering while leveraging sophisticated data structures and algorithmic techniques to achieve remarkable efficiency gains. Its impact has been far-reaching, enabling the application of complete-linkage clustering to domains that were previously inaccessible due to computational constraints.

4.8.1 *Detailed description of the algorithm*

The pseudocode below describes the main steps in the CLINK algorithm.

The algorithm starts by requiring a distance matrix D of size $n \times n$, where n is the number of elements to be clustered. This matrix encapsulates the pairwise distances between all elements. The algorithm initializes an empty min-heap data structure H. This min-heap will be used to retrieve the closest pair of clusters at each iteration efficiently. In a nested loop, the algorithm populates the min-heap H with all unique pairs of elements $\{i, j\}$ and their corresponding distances $D[i, j]$. This initialization step ensures that the min-heap contains the complete set of pairwise distances, facilitating the subsequent merging operations. Then, An array C of size n is initialized to represent the cluster assignments. Initially, each element is assigned to its own cluster, indicated by setting $C[i]$ to i.

The main loop iterates $n - 1$ times, as the algorithm will perform $n - 1$ merging operations to construct the complete hierarchical clustering dendrogram. At each iteration, the closest pair of clusters $\{i, j\}$ and their distance d are extracted from the min-heap H using the ExtractMin operation. The indices p and q of the clusters containing elements i and j, respectively, are determined using the cluster assignments in C. A conditional check is performed to ensure that the clusters p and q are distinct before merging them. This step avoids redundant computations and maintains the integrity of the clustering process. If the clusters p and q are distinct, the algorithm updates the cluster assignments in C by reassigning all elements originally belonging to cluster q to cluster p. This effectively merges the two clusters. After merging the clusters, the algorithm updates the min-heap H to reflect the new distances between the merged cluster and the remaining clusters. For each element m that is not in the merged cluster p, a new distance d' is computed as the maximum distance between m and any element in the merged cluster. This new distance d' and the corresponding cluster pair $\{p, C[m]\}$ are inserted into the min-heap H. Once the main loop completes, the final cluster assignments are stored in the array C. The dendrogram representing the hierarchical clustering can be constructed from these assignments.

The CLINK algorithm leverages the min-heap data structure to efficiently retrieve the closest pair of clusters at each iteration, reducing the overall time complexity to $O(n^2)$. This remarkable feat is achieved through the careful management of cluster assignments and the incremental updates to the min-heap, avoiding the need for costly distance computations between all pairs of clusters at every step.

4.8.2 *Running the algorithm in R*

To illustrate the practical application of the CLINK algorithm, let us consider an example implementation in the R programming language. The R code given in Figure 4.15 demonstrates the usage of the CLINK algorithm for complete-linkage clustering on a sample dataset. In this example, we first load the required "cluster" package, which provides functionality for clustering algorithms in R. We then generate a sample dataset consisting of 100 observations with 5 features, using the "rnorm" function to generate random values from a normal distribution. The "dist" function is used to compute the distance matrix from the sample dataset. This distance matrix serves as the input for the CLINK algorithm. Next, we invoke the "hclust" function, which implements various hierarchical clustering algorithms in R. By setting the "method" parameter to "complete", we specify that we want to perform complete-linkage clustering. The "members" parameter is set to "NULL", indicating that we do not have any a priori information about cluster memberships. Finally, we use the "plot" function to visualize the resulting dendrogram, which provides a graphical representation of the

Require: Distance matrix D of size $n \times n$
Ensure: Hierarchical clustering dendrogram
1: Initialize an empty min-heap H
2: **for** $i \leftarrow 1$ to n **do**
3:    **for** $j \leftarrow i + 1$ to n **do**
4:        Insert $\{i, j, D[i, j]\}$ into H
5:    **end for**
6: **end for**
7: Initialize an array C of size n, representing cluster assignments
8: **for** $i \leftarrow 1$ to n **do**
9:    $C[i] \leftarrow i$
10: **end for**
11: **for** $k \leftarrow 1$ to $n - 1$ **do**
12:    $\{i, j, d\} \leftarrow \text{ExtractMin}(H)$
13:    $p \leftarrow C[i], q \leftarrow C[j]$
14:    **if** $p \neq q$ **then**
15:        **for** $m \leftarrow 1$ to n **do**
16:            **if** $C[m] = q$ **then**
17:                $C[m] \leftarrow p$
18:            **end if**
19:        **end for**
20:        **for** $m \leftarrow 1$ to n **do**
21:            **if** $m \neq p$ and $C[m] \neq p$ **then**
22:                $d' \leftarrow \max\{D[m, i], D[m, j]\}$
23:                Insert $\{p, C[m], d'\}$ into H
24:            **end if**
25:        **end for**
26:    **end if**
27: **end for**
28: Construct the dendrogram from the cluster assignments in C

Fig. 4.15: CLINK algorithm for complete-linkage clustering.

hierarchical clustering obtained by the CLINK algorithm. This example demonstrates the ease of use and the broad applicability of the CLINK algorithm in the R programming environment. With a few lines of code, we can leverage the power of this efficient algorithm.

Listing 4.11: Using CLINK algorithm to cluster dataset

```
1
2 # Load required packages
3 library(cluster)
4
5 # Generate a sample dataset
6 set.seed(123)
7 data <- matrix(rnorm(100 * 5), ncol = 5)
8
9 # Compute the distance matrix
10 dist_matrix <- dist(data)
11
12 # Perform complete-linkage clustering using CLINK
13 fit <- hclust(dist_matrix, method = "complete", members =
       NULL)
14
15 # Plot the dendrogram
16 plot(fit, main = "Complete-Linkage Clustering Dendrogram")
```

4.9   Unweighted Pair Group Method with Arithmetic Mean (UPGMA)

The Unweighted Pair Group Method with Arithmetic Mean (UPGMA) is a simple, bottom-up hierarchical clustering method. It is widely used in various fields, such as bioinformatics and ecology, for constructing phylogenetic trees. UPGMA assumes a constant rate of evolution, making it an ultrametric method. This assumption is crucial as it influences the interpretation of the resulting dendrogram.

The UPGMA algorithm operates by iteratively merging the closest pair of clusters until only a single cluster remains. At each iteration, it recalculates the distances between the newly formed cluster and all remaining clusters using the average distance. The process results in a rooted tree that reflects the hierarchical structure of the data.

The UPGMA algorithm is described in the pseudocode given in Figure 4.16.

Let's go through the pseudo-code line by line:

(1) We start by initializing each data point as its own cluster.

(2) The distance matrix, which contains pairwise distances between clusters, is computed.

(3) We enter a loop that continues until all data points are merged into a single cluster.

1: Initialize n clusters, each containing one element.
2: Compute the initial distance matrix.
3: **while** more than one cluster exists **do**
4:   Find the pair of clusters C_i and C_j with the smallest distance.
5:   Merge C_i and C_j to form a new cluster C_{ij}.
6:   Update the distance matrix to reflect the new cluster.
7: **end while**
8:
9: **return** the hierarchical clustering.

Fig. 4.16: UPGMA algorithm.

(4) Within the loop, we identify the two clusters with the smallest distance.
(5) These two clusters are merged to form a new cluster.
(6) The distance matrix is updated to include the new cluster and exclude the merged clusters.

4.9.1 *Running the algorithm in R*

To implement UPGMA in R, we can use the "hclust" function from the "stats" package. We first prepare the data and then compute the distance matrix using the dist function. Now, we apply UPGMA by using the "hclust" function with the method set to "average". Finally, we visualize the resulting dendrogram.

Listing 4.12: Using UPGMA hierarchical clustering

```
1 # Load necessary library
2 library(stats)
3
4 # Prepare data
5 data <- matrix(rnorm(30), nrow=10, ncol=3)
6
7 # Compute distance matrix
8 dist_matrix <- dist(data)
9
10 # Apply UPGMA
11 hc <- hclust(dist_matrix, method = "average")
12
13 # Plot dendrogram
14 plot(hc, main="Dendrogram of UPGMA Clustering", xlab="Sample
       Index", sub="", ylab="Height")
```

4.9.2   *Conclusions, advantages, limitations,*
        and future research

UPGMA is a foundational method in hierarchical clustering with several notable advantages and limitations. Here, we summarize its key points. The advantages of the algorithm include

(1) **Simplicity:** UPGMA is straightforward to understand and implement.
(2) **Efficiency:** It is computationally efficient for small to moderate-sized datasets.
(3) **Interpretability:** The resulting dendrogram provides a clear visual representation of the data's hierarchical structure.

UPGMA also has several limitations, such as:

(1) **Ultrametric assumption:** UPGMA assumes a constant rate of evolution, which may not hold in all cases.
(2) **Sensitivity to noise:** The algorithm can be sensitive to outliers and noisy data.
(3) **Fixed number of clusters:** UPGMA does not provide a natural way to determine the optimal number of clusters.

Future research could explore:

(1) **Relaxing the ultrametric assumption:** Developing methods that do not require the assumption of a constant rate of evolution.
(2) **Robustness improvements:** Enhancing UPGMA to be more robust to noise and outliers.
(3) **Integration with other methods:** Combining UPGMA with other clustering methods to improve accuracy and interoperability.

In summary, UPGMA is a valuable tool in the data scientist's toolkit, offering a balance of simplicity and interpretability. While it has its limitations, ongoing research and methodological advancements continue to enhance its applicability and robustness.

4.10   WPGMA Algorithm

The Weighted Pair Group Method with Arithmetic Mean (WPGMA) is an agglomerative hierarchical clustering algorithm, a method generally attributed to Sokal and Michener. This method constructs a rooted tree, or dendrogram, to represent the hierarchical relationships between data points based on a pairwise distance matrix. The dendrogram provides a visual representation of the nested clusters formed during the algorithm's execution. WPGMA is closely related to the unweighted variant, UPGMA, but differs in the way it averages distances.

In the WPGMA algorithm, the distance between clusters is computed as the arithmetic mean of the distances between all pairs of elements, one from each cluster. This method assumes a constant rate of evolution, meaning that it produces an ultrametric tree where all branch tips are equidistant from the root. This ultrametricity assumption is often referred to as the molecular clock in the context of DNA, RNA, and protein data.

The WPGMA algorithm follows a straightforward sequence of steps to merge clusters based on their pairwise distances. Figure 4.17 gives a detailed description of the pseudo-code.

1: Initialize each element as a separate cluster.
2: **while** more than one cluster remains **do**
3:    Identify the pair of clusters (i, j) with the smallest distance.
4:    Merge clusters (i, j) into a new cluster $(i \cup j)$.
5:    Update the distance matrix: for each cluster k, compute $d_{(i \cup j),k} = \frac{d_{i,k} + d_{j,k}}{2}$.
6: **end while**
7: Construct the final dendrogram with the clusters.

Fig. 4.17: WPGMA algorithm.

The algorithm starts by treating each element as its own cluster. Then, it iteratively identifies the two clusters that are closest together, merging them into a single cluster. After each merge, the distance matrix is updated to reflect the new distances between the merged cluster and all remaining clusters. Specifically, the distance between the new cluster and any other cluster is computed as the arithmetic mean of the distances between each original cluster and the other cluster. This process is repeated until all elements are combined into a single cluster, resulting in the construction of the final dendrogram.

To implement the WPGMA algorithm in R, we can utilize several functions from the base R and external packages. In this example, we start with a sample distance matrix representing the pairwise distances between five elements. We then load the necessary libraries and convert the distance matrix into a `dist` object, which is the required format for the clustering

Listing 4.13: Using WPGMA hierarchical clustering

```
1  # Sample distance matrix
2  distance_matrix <- matrix(c(
3    0, 18, 21, 31, 23,
4    18, 0, 30, 34, 20,
5    20, 30, 0, 28, 40,
6    31, 34, 28, 0, 43,
7    23, 21, 40, 43, 0
8  ), nrow = 5, byrow = TRUE)
9  rownames(distance_matrix) <- colnames(distance_matrix) <- c(
      'a', 'b', 'c', 'd', 'e')
10
11 # Load necessary libraries
12 library(cluster)
13
14 # Convert distance matrix to a dist object
15 dist_object <- as.dist(distance_matrix)
16
17 # Perform hierarchical clustering using WPGMA method
18 wpgma_clustering <- agnes(dist_object, method = "weighted")
19
20 # Plot the dendrogram
21 plot(wpgma_clustering, which.plots = 2, main = "WPGMA
       Dendrogram")
```

functions in R. The `agnes` function from the `cluster` package performs hierarchical clustering using the WPGMA method, specified by the `method = "weighted"` argument. Finally, we plot the resulting dendrogram to visualize the hierarchical relationships.

The WPGMA algorithm is a simple yet effective method for hierarchical clustering, producing rooted dendrograms that reflect the nested structure of the data. Here are the main points regarding its usage:

The advantages of the WPGMA algorithm include

(1) The algorithm is straightforward to implement and understand.
(2) It produces a clear and interpretable dendrogram.
(3) Suitable for datasets where the constant-rate assumption (ultrametricity) holds.

WPGMA also has several limitations, such as:

(1) The constant-rate assumption may not be valid for all datasets, leading to inaccurate clustering.
(2) It can be computationally intensive for large datasets due to the repeated distance matrix updates.

(3) The method is sensitive to the initial distance matrix, and small changes can significantly affect the resulting dendrogram.

Future research could explore:

(1) Developing methods to relax the constant-rate assumption while retaining the simplicity of WPGMA.
(2) Improving computational efficiency for handling larger datasets.
(3) Exploring the integration of WPGMA with other clustering methods to enhance its robustness and accuracy.

In summary, while WPGMA is a valuable tool for hierarchical clustering, its assumptions and computational demands necessitate careful consideration. Ongoing research aims to address these challenges, making the algorithm more flexible and scalable for diverse applications.

4.11 Comparing the Clustering of SLINK, CLINK, UPGMA and WPGMA

In order to compare the clustering results of various hierarchical clustering methods, let us consider a dataset consisting of five elements (a, b, c, d, e) and the following matrix of pairwise distances between them:

	a	b	c	d	e
a	0	18	21	31	23
b	18	0	30	34	20
c	21	30	0	28	40
d	31	34	28	0	43
e	23	20	40	43	0

One of the commonly used methods is complete linkage clustering, also known as CLINK. This method overcomes a significant drawback of the single linkage clustering (SLINK) method — the chaining phenomenon. In the SLINK method, clusters formed via single linkage clustering may be forced together due to single elements being close to each other, even though many of the elements in each cluster may be very distant from each other.

The CLINK method, on the other hand, defines the distance between two clusters as the maximum distance between any pair of elements, one from each cluster. This approach tends to find compact clusters of approximately equal diameters, making it less susceptible to the chaining effect observed in SLINK. CLINK produces more robust and well-separated clusters, especially when the data contains outliers or noise.

In the previous sections, we presented another popular method known as the Weighted Pair Group Method with Arithmetic Mean (WPGMA), which calculates the distance between two clusters as the arithmetic mean of the distances between all pairs of elements, one from each cluster. The Unweighted Pair Group Method with Arithmetic Mean (UPGMA) is a variant of WPGMA, where it assumes that all clusters have equal weights or sizes.

For the given small dataset, WPGMA and UPGMA are expected to produce similar clustering results, as the arithmetic mean of the distances between elements will be comparable due to the limited number of data points. Both methods aim to strike a balance between the compactness and separation of clusters, falling between the extremes of SLINK and CLINK.

To accurately compare the clustering results, it is recommended to visualize the dendrograms obtained from each method and analyze the results. Figure 4.18 shows the resulting dendrograms of each method on the given

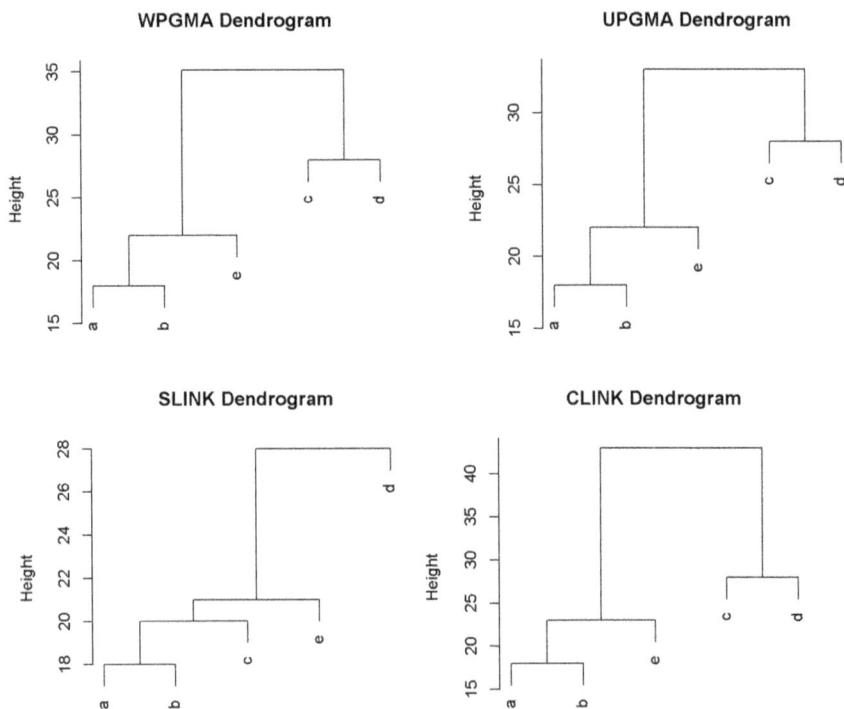

Fig. 4.18: Comparing dendrograms from different methods.

dataset. The choice of the clustering method depends on the specific characteristics of the data and the desired properties of the clusters, such as compactness, separation, or resistance to outliers.

4.12 Sequential Agglomerative Hierarchical Non-overlapping (SAHN) Algorithm

The SAHN (Sequential Agglomerative Hierarchical Non-overlapping) algorithm is a fundamental hierarchical clustering method that produces a hierarchy of nested clusters by merging smaller clusters into larger ones. This chapter provides a comprehensive overview of SAHN clustering.

The key idea behind SAHN is to start with each data point in its own singleton cluster and then repeatedly merge the two closest clusters until all points belong to one cluster. The following sentence presents the major steps involved:

There are four major steps in the SAHN algorithm: 1) Initialize by assigning each data point to its own cluster, 2) Compute the proximity matrix containing the distance between all pairs of clusters, 3) Find the closest pair of clusters and merge them into a new cluster, 4) Update the proximity matrix to reflect this merge, and repeat steps 3-4 until only one cluster remains.

The pseudocode for SAHN clustering is presented in Figure 4.19. The algorithm takes an $n \times p$ data matrix X as input, where n is the number of data points and p is the number of features/dimensions. It initializes n singleton clusters, where each cluster C_i contains just the i-th data point x_i. Then, it computes the $n \times n$ proximity matrix D, where D_{ij} contains

Require: X is the $n \times p$ data matrix
1: Initialize n singleton clusters $C_i = \{x_i\}$ for $i = 1 \ldots n$
2: Compute $n \times n$ proximity matrix D containing distances between all cluster pairs
3: **for** $i = n$ to 2 **do**
4:    Find closest pair of clusters C_r, C_s in D
5:    Merge C_r and C_s into a new cluster C_{rs}
6:    Update D to reflect C_{rs} instead of C_r, C_s'
7: **end for**
8:
9: **return** Hierarchy of nested clusters

Fig. 4.19: SAHN clustering.

the distance/dissimilarity between clusters C_i and C_j. Common choices for the distance metric include Euclidean distance, Manhattan distance, correlation distance etc. It then enters a loop that runs $(n-1)$ times. In each iteration: A) It finds the closest pair of clusters C_r, C_s in the proximity matrix D; B) It merges C_r and C_s into a new combined cluster C_{rs}; C) It updates D to reflect the new cluster C_{rs} instead of the old C_r and C_s. The distances between C_{rs} and the remaining clusters must be recomputed.

After $(n-1)$ iterations, all points are merged into one cluster. Finally, it returns the hierarchy of nested clusters produced along the way. The key outcome of SAHN is this hierarchy, which can be visualized as a dendrogram and interpreted at different levels to obtain different clusterings.

The SAHN algorithm is a simple yet powerful method for hierarchical clustering. Some of its main advantages are:

- It does not require specifying the number of clusters a priori.
- It can identify clusters of varying shapes, sizes and densities.
- The hierarchy provides insight into similarities between data at multiple scales.

However, it also has some important limitations:

- With n data points, it has a computational complexity of $\Theta(n^2)$ in both memory and time
- The hierarchy is sensitive to noise and outliers
- The merging process is inflexible and cannot be undone

Looking ahead, some exciting future research directions include:

- Extending SAHN to cluster data streams and massive datasets using approximations
- Developing more robust hierarchical methods that can handle outliers and noise
- Hybridizing SAHN with flat clustering techniques like k-means

Overall, SAHN remains a widely used and highly interpretable clustering technique that every data scientist should know.

4.13　The CURE Clustering Algorithm

The CURE (Clustering Using Representatives) algorithm is a robust clustering method designed to handle large datasets and overcome the limitations of traditional clustering algorithms [Guha *et al.* (1998b)]. CURE achieves this by representing clusters with a fixed number of well-scattered points and employing a combination of random sampling and partitioning to cluster large datasets efficiently.

CURE's approach to clustering is particularly effective in dealing with outliers and identifying clusters of arbitrary shapes and sizes, making it a versatile tool in the data analyst's toolkit.

CURE clustering algorithm can be broken down into the following steps:

(1) **Random sampling:** Select a random sample of the dataset to reduce computational complexity.
(2) **Partitioning and partial clustering:** Partition the random sample and cluster each partition partially.
(3) **Representative points selection:** For each cluster, select a fixed number of representative points.
(4) **Shrinking towards the mean:** Shrink these representative points towards the mean of the cluster.
(5) **Hierarchical clustering:** Perform hierarchical clustering on the representative points.

The pseudo-code for the CURE algorithm is given in Figure 4.20.

Require: Data set D, number of clusters k, number of representative points c, shrinking factor α
Ensure: Clusters of D
1: Select a random sample S from D
2: Partition S into p partitions
3: **for** each partition P_i **do**
4:    Partially cluster P_i into k_i clusters
5: **end for**
6: Merge all partial clusters into a single set C
7: Select c representative points for each cluster in C
8: **for** each representative point r **do**
9:    Shrink r towards the mean of its cluster by α
10: **end for**
11: Perform hierarchical clustering on the representative points
12: Assign data points in D to the nearest cluster
13:
14: **return** final clusters

Fig. 4.20: CURE clustering algorithm.

Now, let's break down each step of the algorithm in detail.

4.13.1   *Random sampling*

Random sampling is employed to reduce the size of the dataset, making the clustering process more efficient. By selecting a representative subset of the data, we ensure that the computational cost is manageable while still capturing the overall structure of the dataset.

4.13.2   *Partitioning and partial clustering*

The random sample is partitioned into smaller subsets, and each subset is partially clustered. This step helps in managing large datasets by breaking down the clustering task into smaller, more manageable pieces.

4.13.3   *Representative points selection*

A fixed number of representative points are selected for each cluster obtained from the partial clustering step. These points are chosen to be well-scattered within the cluster, capturing the diversity and structure of the cluster effectively.

4.13.4   *Shrinking towards the mean*

Each representative point is then shrunk towards the mean of its cluster by a specified shrinking factor. This step helps mitigate the effects of outliers and ensures that the representative points are more central to their respective clusters.

4.13.5   *Hierarchical clustering*

Finally, hierarchical clustering is performed on the representative points to merge them into the final clusters. This step ensures that clusters of arbitrary shapes and sizes can be effectively identified.

4.13.6   *Conclusions, advantages, limitations, and future research*

The CURE algorithm offers several advantages:

- **Robustness to outliers:** By shrinking representative points towards the mean, CURE effectively mitigates the impact of outliers.
- **Scalability:** The use of random sampling and partitioning allows CURE to handle large datasets efficiently.
- **Flexibility:** CURE can identify clusters of arbitrary shapes and sizes.

However, there are also some limitations:

- **Complexity:** The algorithm involves several steps, making it more complex to implement and tune compared to simpler clustering algorithms.
- **Parameter sensitivity:** The performance of CURE can be sensitive to the choice of parameters, such as the number of representative points and the shrinking factor.

Future research directions could focus on:

- **Parameter optimization:** Developing methods for automatic parameter tuning to improve the algorithm's performance and ease of use.
- **Parallel processing:** Implementing parallel versions of the algorithm to enhance its scalability further.
- **Hybrid approaches:** Combining CURE with other clustering methods to leverage their strengths and overcome CURE's limitations.

In summary, the CURE clustering algorithm is a powerful tool for clustering large datasets, offering robustness and flexibility. Its effectiveness in dealing with outliers and identifying complex cluster structures makes it a valuable addition to the data analyst's toolkit.

4.14 Nearest-neighbor Chain Algorithm

The nearest-neighbor chain algorithm is a powerful method used in hierarchical clustering. This algorithm efficiently constructs a hierarchical clustering by iteratively merging clusters based on the nearest-neighbor criterion. Unlike traditional hierarchical methods, which can be computationally expensive, the nearest-neighbor chain algorithm reduces computational complexity and accelerates the clustering process. The main concept revolves around the nearest-neighbor relationship between clusters, forming a chain that dynamically grows until a merge is necessitated.

4.14.1 *Detailed description of the algorithm*

To understand the nearest-neighbor chain algorithm, let's delve into its step-by-step procedure. We start by initializing each data point as its own cluster. Then, we iteratively find and merge the nearest clusters until only a single cluster remains.

In essence, the algorithm operates as follows: Initially, each data point is treated as an individual cluster. An empty chain is created to keep track of

Require: Data points $\{x_1, x_2, \ldots, x_n\}$
Ensure: Hierarchical clustering
 1: Initialize each data point as a separate cluster
 2: Initialize an empty chain
 3: **while** more than one cluster remains **do**
 4:   **if** chain is empty **then**
 5:     Select an arbitrary cluster and add it to the chain
 6:   **end if**
 7:   Let C be the last cluster in the chain
 8:   Find the nearest neighbor C' of C
 9:   **if** C' is the previous cluster in the chain **then**
 10:     Merge C and C' to form a new cluster
 11:     Remove C and C' from the chain
 12:   **else**
 13:     Add C' to the chain
 14:   **end if**
 15: **end while**
 16: **return** The sequence of merges as a hierarchical clustering

Fig. 4.21: Nearest-neighbor chain algorithm.

the nearest neighbors. If the chain is empty, a cluster is arbitrarily selected and added to it. The algorithm then identifies the nearest neighbor of the last cluster in the chain. If this neighbor is already in the chain, the two clusters are merged, and the process continues. Otherwise, the neighbor is added to the chain. This cycle repeats until only one cluster remains, resulting in the hierarchical clustering of the data (Figure 4.21).

4.14.2 *Conclusions, advantages, limitations, and future research*

The nearest-neighbor chain algorithm offers several benefits for hierarchical clustering:

- **Efficiency:** By reducing the number of distance calculations, the algorithm speeds up the clustering process.
- **Simplicity:** The algorithm is straightforward to implement and understand.
- **Dynamic chaining:** The chain mechanism dynamically adjusts, leading to more accurate clustering.

However, there are limitations to consider:

- **Initial selection:** The arbitrary selection of the initial cluster can influence the resulting clustering.
- **Scalability:** While more efficient than traditional methods, the algorithm may still face scalability issues with very large datasets.

Future research can explore:

- **Improved initialization:** Developing strategies for better initial cluster selection.
- **Parallel processing:** Leveraging parallel computing to speed up the algorithm further.
- **Robustness:** Enhancing the algorithm to handle noisy data more effectively.

The nearest-neighbor chain algorithm stands out as a practical and efficient approach to hierarchical clustering. Its ability to dynamically adjust the clustering process offers significant advantages, making it a valuable tool for data scientists and researchers alike.

Chapter 5

Clustering Visualization

5.1 Introduction

Clustering visualization is an important way to gain insight into how clustering methods work and to compare the results of different methods. This chapter will present several common clustering visualization techniques and R packages.

5.2 Using Built-in Plot Function

We begin by reviewing simple approaches for plotting clustering results. Consider 12 points randomly selected from the Iris dataset and projected into two dimensions: Petal.Width and Petal.Length. As illustrated in List 5.1, we can use the internal *plot* function to display these points. The resulting graph is presented in Figure 5.1.

Listing 5.1: Using *plot* function to present the Iris dataset

```
1 set.seed(1123)
2 iris2dsample <- iris[sample(nrow(iris),12),3:4]
3 plot(iris2dsample,pch=19,cex=2)
```

In List 5.2 we cluster the 12 points into three clusters using the k-means algorithm. Then, we are displaying these points such that each cluster is represented using a different symbol (square, circle and triangle) (Figure 5.2). This is done in Line 4 by setting the *pch* parameter of *plot* function to the cluster number. In Line 5, we display the clusters' centers using the corresponding symbol but using its filled version in order to differentiate the centers from the actual points. For example, the points of the first clus-

135

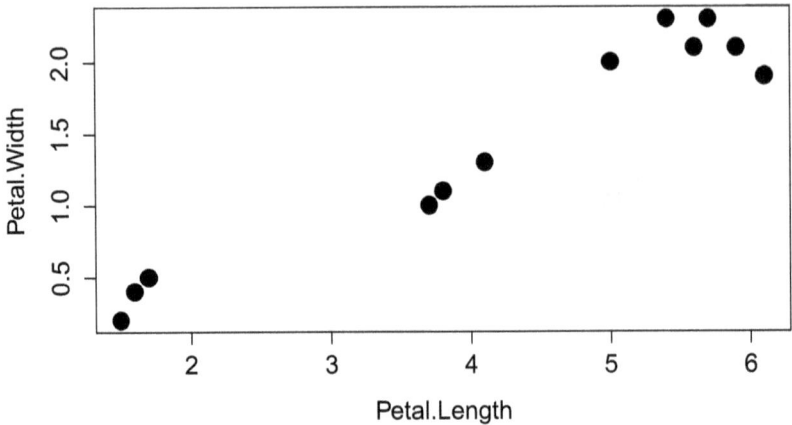

Fig. 5.1: Graph of 12 points.

ter are represented as an empty square, and the center of the first cluster is represented as a filled square.

On Lines 7-10, the original points are connected to the corresponding centers using the *segment* function. Specifically, in Line 7, we are using the *for* loop to iterate over the three clusters. In Line 8, we are using the *segment* function to draw line segments between pairs of points (in this case, the original point and its corresponding center).

Listing 5.2: Clustering 12 points into three clusters and plotting the centroids and the points assignment

```
1  set.seed(1123)
2  iris2dsample <- iris[sample(nrow(iris),12),3:4]
3  cmeans=kmeans(iris2dsample, 3)
4  plot(iris2dsample, pch=cmeans$cluster-1,xlab="",ylab="",cex
     =1.5,col=1)
5  points(cmeans$centers, col = 1, pch = 14+1:3, cex = 1.5)
6
7  for (i in 1:length(cmeans$centers[,1])) {
8  segments( iris2dsample[cmeans$cluster==i,][,1], iris2dsample
     [cmeans$cluster==i,][,2],
9            cmeans$centers[i,1], cmeans$centers[i,2])
10 }
```

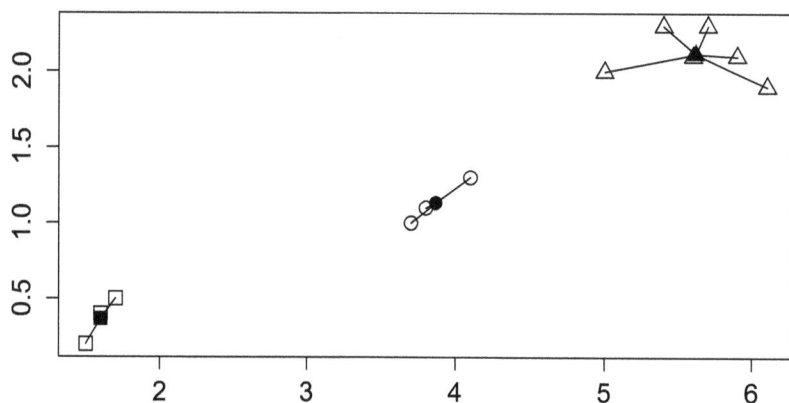

Fig. 5.2: Clustering results using symbols and connected lines.

5.3 The Clusplot Function

The *clusplot* function in the *cluster* package is an essential function for drawing two-dimension clustering results. It gets as an imput a "partition" object, that is created by one of the clustering functions such as: *pam*, *clara*, or *fanny* (Figure 5.3).

Listing 5.3: Using *clustplot* to plot the clusters assignments

```
1 library(cluster)
2
3 set.seed(1123)
4 iris2dsample <- iris[sample(nrow(iris),12),3:4]
5 cmeans=pam(iris2dsample, 3)
6 clusplot(cmeans,color=TRUE,shade=TRUE,lines=0)
```

5.4 FlexClust Package

The *FlexClust* package provides several functions for plotting the clustering results. The *barchart* function displays a two-dimension array of bars. The columns correspond to the clusters, and the rows refer to the variables. List 5.4 illustrates the usage of *barchart* function. In Line 1, we load the *FlexClust* package. In Line 2, we are performing k-means clustering using

clusplot(pam(x = iris2dsample, k = 3))

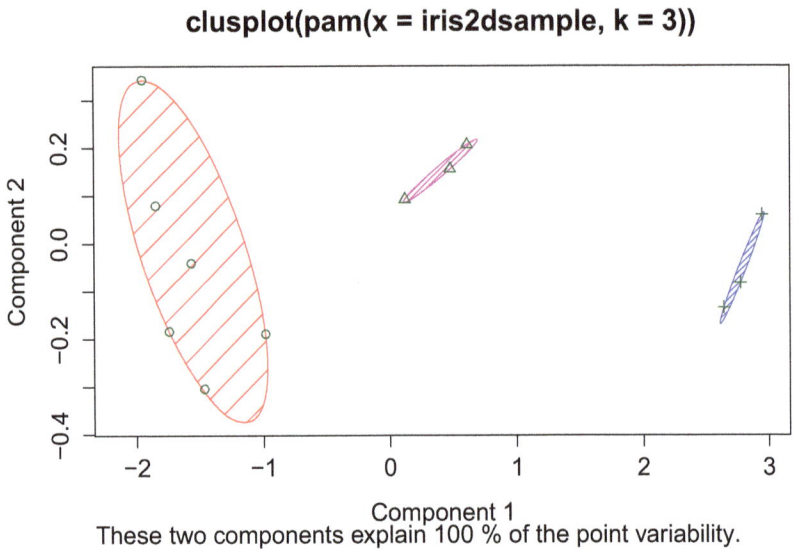

Component 1
These two components explain 100 % of the point variability.

Fig. 5.3: Clustering results using *clustplot* function.

the *cclust* function. By setting the *method* of *cclust* function, it is possible to use two other clustering algorithms: If the value *"hardcl"* is used, then hard competitive clustering is performed. It works by randomly selecting a point and moving the closest center towards that point [Ripley (1996)]. If the value *"neuralgas"* is used, then the neural gas algorithm is used [Martinetz *et al.* (1993)]. Similarly to hard competitive clustering, the closest centroid is moved but in addition also the second closest centroid is moved. In Line 3, the bar chart is plotted as illustrated in Figure 5.4.

Listing 5.4: Using *barchart* function to plot the variables values in each cluster

```
1 library(flexclust)
2 cl <- cclust(iris[,-5], k=3, save.data=TRUE)
3 barchart(cl)
```

The *bwplot* function draws a boxplot for each variable in each cluster in comparison with boxplot for the complete sample. List 5.5 illustrates the usage of the *bwplot* function and the resulted graph is presented in Figure 5.5.

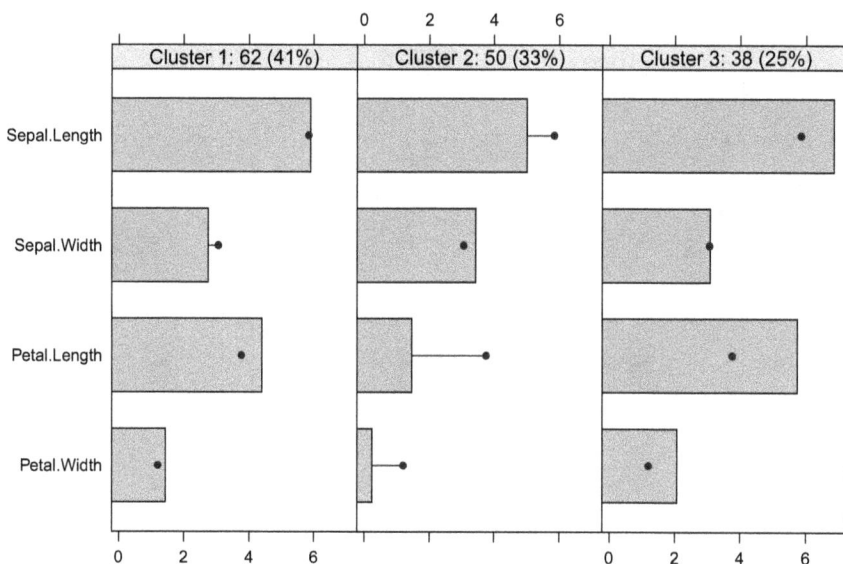

Fig. 5.4: Illustrating *barchart* function.

Listing 5.5: Using *bwplot* function to plot the variables values in each cluster

```
1 library(flexclust)
2 cl <- cclust(iris[,-5], k=3, save.data=TRUE)
3 bwplot(cl)
```

The *plot* function draws the neighborhood graph of a cluster solution together with projected points. Similarly *image* plots of cluster segments overlaid by a neighborhood graph. In both functions, in addition to the clustering object, the user can determine which two dimensions will be presented in the graph by setting the *which* parameter as illustrated in List 5.6. In this case we are using the third and fourth dimensions for drawing the graph. The resulting graph is presented in Figure 5.6.

Listing 5.6: Using *plot* function to draw the clusters

```
1 library(flexclust)
2 cl <- cclust(iris[,-5], k=3, save.data=TRUE)
3 plot(cl,which=c(3,4))
```

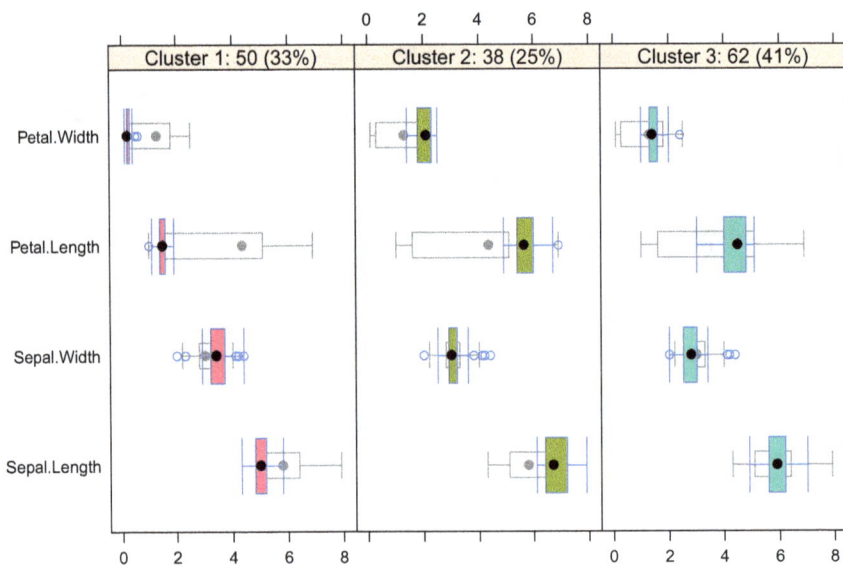

Fig. 5.5: Illustrating *bwplot* function.

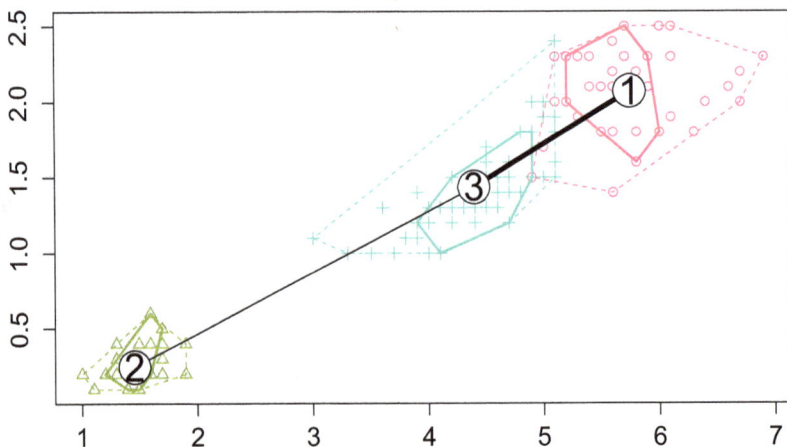

Fig. 5.6: Illustrating *plot* function.

The function *shadow* calculates the shadow value of each point. The shadow value is defined as twice the distance to the closest centroid divided by the sum of distances to the closest and second-closest centroid. If most of the points in a cluster have low shadow values, it indicates that the cluster is well separated from all other clusters [Leisch (2010)].

The silhouette value is defined as the scaled difference between the average dissimilarity of a point to all points in its own cluster to the smallest average dissimilarity to the points of a different cluster. Well-separated clusters should have high silhouette values. The *Silhouette* function is used to get these values, as illustrated in List 5.7.

Listing 5.7: Illustrating the usage of *shadow* and *Silhouette* functions.

```
1 library(flexclust)
2 set.seed(1123)
3 cl <- cclust(iris[,-5], k=3, save.data=TRUE)
4 ## high shadow values indicate clusters with *bad*
     separation
5 plot(shadow(cl))
6 ## high Silhouette values indicate clusters with *good*
     separation
7 plot(Silhouette(cl))
```

A shadow star graph depicts the distribution of shadow values. The centroids are represented as nodes in a graph, which are connected by stripe plots of shadow values. The graph connects two nodes by a line if at least one point has these two centroids as the closest centroid and second closest centroid. The width of the line is proportional to the sum of shadow values of all points having these two as being closest and the second closest. The shadow star graph can be used to assess which clusters are close to each other. For well-separated clusters, the shadow values concentrate their mass close to the centroids. List 5.8 demonstrates the usage of the shadow star graph using violin plot as illustrated in Figure 5.7

Listing 5.8: Using *plot* function to draw the clusters

```
1 library(flexclust)
2 set.seed(1123)
3 cl <- cclust(iris[,-5], k=3, save.data=TRUE)
4 shadowStars(cl,which=c(3,4), panel=panelShadowViolin)
```

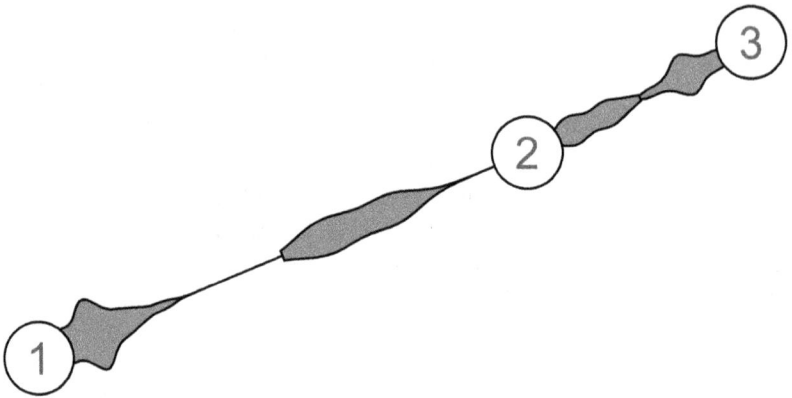

Fig. 5.7: Illustrating shadow star graph.

5.5 Dendrogram

A dendrogram is the most popular diagram for illustrating the arrangement
of the clusters produced by hierarchical clustering. A dendrogram is a tree
structure (in Greek dendro is a "tree") that represents the nested grouping
of objects (in an agglomerative method) or the nested partitioning (in a
divisive method). Each branch in the tree is called a *clade*. The terminal
end of each clade is called a leaf. The leaves of the dendrogram (usually the
lowest nodes in the tree) represent the individual instances. The remaining
nodes represent the clusters to which the instances are assigned. Clades
can have just one leaf, or they can have more than one. The root of the
dendrogram (the node located at the top of the graph) represents the universal cluster that consists of all instances in the dataset. The height of
each node in the graph refers to the distance between its two children. In
addition, the dendrogram shows the similarity levels at which groupings
change. A certain clustering of the data can be obtained by cutting the
dendrogram at the desired similarity level.

Let us go back to Figure 4.8, which shows a dendrogram for 12 randomly
selected flowers from the Iris dataset. At height=0, all twelve flowers are
singleton clusters. At height=0.1, flowers 141 and 78 are joined together.
Then, at height=0.141, flowers 105 and 138 are joined together. The process
continues until all points are merged at height=2.061 into one large cluster.

The class *dendrogram* in package *stats* provides general functions for
plotting or cutting dendrograms. List 5.9 illustrates the usage of the dendrogram package. We begin by selecting a sample of 12 flowers from the

Iris dataset (line 5). In line 8, the data is clustered using the function hclust which implements a hierarchical clustering algorithm. In line 11, the primary dendrogram object is created using the obtained hierarchical clustering. The dendrogram is plotted in line 14 using the default arguments.

Listing 5.9: Using *dendrogram* class to draw the clusters

```
1  require(graphics);
2  require(utils)
3
4  set.seed(1123)
5  iris2dsample <- iris[sample(nrow(iris),12),3:4]
6
7  # clustering the data
8  hc <- hclust(dist(iris2dsample), "ave")
9
10 # creating the basic dendrogram
11 dend <- as.dendrogram(hc)
12
13 ## ploting the dendrogram
14 plot(dend)
15
16 ## ploting the dendrogram using "triangle" type and showing
      the inner nodes:
17 plot(dend, nodePar = list(pch = c(1,NA), cex = 0.8, lab.cex
      = 0.8), type = "tr", center = TRUE)
18
19 ## presenting dendrogram in a text format
20 str(dend)
21
22 ## Using only the first three sub-levels
23 str(dend, max = 3)
24
25 ## Cutting the tree in the highet of 0.4
26 dendcut <- cut(dend, h = 0.4)
27
28 ## Ploting the upper part of the dendrogram
29 plot(dendcut$upper, type="tr")
30
31 ## Ploting the lower part of the dendrogram, note that
32 ## $lower is NOT a dendrogram, but a list of items
33 plot(dendcut$lower[[3]], nodePar = list(col = 4), type = "tr
      ")
```

In line 17, we plot the dendrogram again, this time using the triangle type. Note that two types of dendrograms are supported: "rectangle" (the default) and "triangle". The results are presented in Figure 5.8. In line 19, the dendrogram is presented in text format. The obtained output is presented in List 5.10.

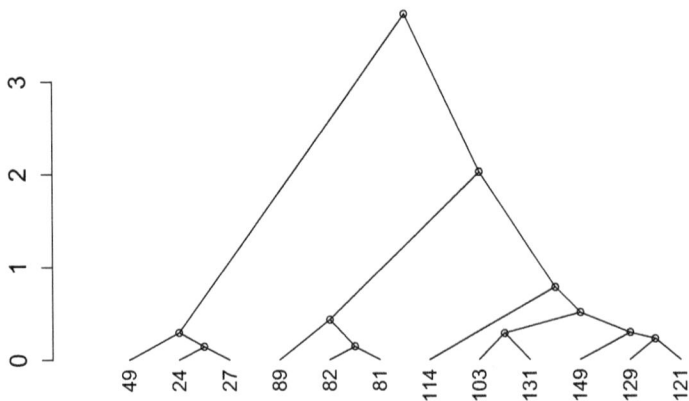

Fig. 5.8: Illustrating dendrogram using triangle type.

Listing 5.10: Using *str* function to print the dendrogram in a text format

```
 1  --[dendrogram w/ 2 branches and 12 members at h = 3.72]
 2  |--[dendrogram w/ 2 branches and 3 members at h = 0.292]
 3  |   |--leaf "49"
 4  |   '--[dendrogram w/ 2 branches and 2 members at h =
        0.141]
 5  |       |--leaf "24"
 6  |       '--leaf "27"
 7  '--[dendrogram w/ 2 branches and 9 members at h = 2.02]
 8      |--[dendrogram w/ 2 branches and 3 members at h = 0.43]
 9      |   |--leaf "89"
10      |   '--[dendrogram w/ 2 branches and 2 members at h =
            0.141]
11      |       |--leaf "82"
12      |       '--leaf "81"
13      '--[dendrogram w/ 2 branches and 6 members at h =
            0.776]
14          |--leaf "114"
15          '--[dendrogram w/ 2 branches and 5 members at h =
                0.505]
16              |--[dendrogram w/ 2 branches and 2 members at h =
                    0.283]
17              |   |--leaf "103"
18              |   '--leaf "131"
19              '--[dendrogram w/ 2 branches and 3 members at h =
                    0.291]
20                  |--leaf "149"
21                  '--[dendrogram w/ 2 branches and 2 members at
                        h = 0.224]
22                      |--leaf "129"
23                      '--leaf "121"
```

In line 22, we textually display the dendrogram using only the first three sub-levels. In line 26, we illustrate how to cut the dendrogram at a certain height. (*h=0.4* in this case). In line 29, the upper part of the cut dendrogram is plotted, and in line 33, the lower part is plotted.

The *dendextend* package enables us to extend, manipulate and compare dendrograms. For example, the *color labels* function lets us selectively color branches and labels in the dendrogram. List 5.11 illustrates the usage of the *color labels* function. The coloring manipulation is performed in line 15. The *col* parameter specifies the color that will be used, and the *labels* parameter indicates which nodes will be colored. In addition to the labels' colors, the *dendextend* package can be used to manipulate many other features of the dendrogram, such as: The labels' size, the nodes' point type, size and color, the branches' color, line type and width.

Listing 5.11: Using *color labels* function for manipulating a dendrogram

```
1  require(graphics);
2  require(utils)
3  require(dendextend)
4
5  set.seed(1123)
6  iris2dsample <- iris[sample(nrow(iris),12),3:4]
7
8  # clustering the data
9  hc <- hclust(dist(iris2dsample), "ave")
10
11 # creating the basic dendrogram
12 dend <- as.dendrogram(hc)
13
14 # coloring some labels, based on label names:
15 dend=color_labels(dend,col = "red", labels = labels(dend)[c
      (4,5)])
16
17 ## ploting the dendrogram with selective coloring of
      branches AND labels :)
18 plot(dend)
```

5.5.1 *Comparing a pair of dendrograms*

A tanglegram is a pair of trees with the same set of leaves (data points in this case) plotted side by side with matching leaves in the two trees connected by a line. Tanglegram is frequently used for visually comparing two dendrograms. List 5.12 illustrates how tanglegram is created using the *dendextend* package. In lines 9-10, two clusterings are created.

Listing 5.12: Using tanglegram for comparing a pair of dendrograms

```
1 require(graphics);
2 require(utils)
3 require(dendextend)
4
5 set.seed(1123)
6 iris2dsample <- iris[sample(nrow(iris),12),3:4]
7
8 # clustering the data
9 hc1 <- hclust(dist(iris2dsample), "ave")
10 hc2 <- hclust(dist(iris2dsample), "com")
11 tanglegram(hc1 , hc2)
```

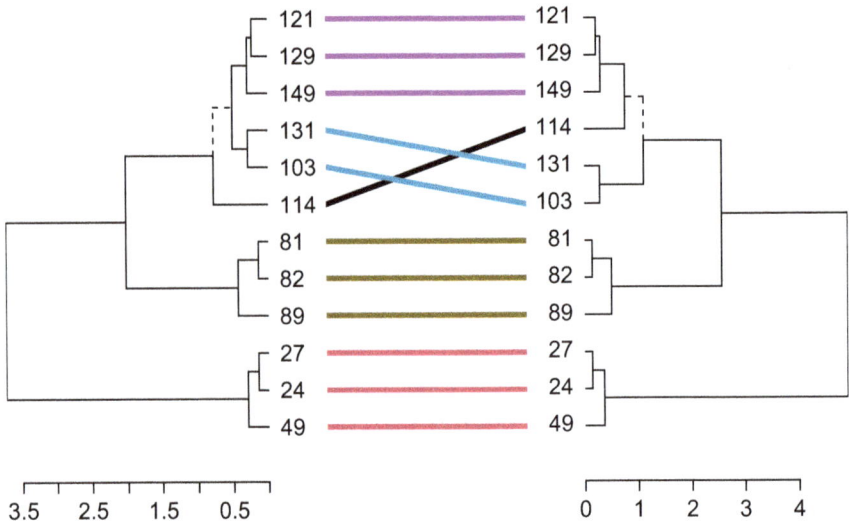

Fig. 5.9: Illustrating tanglegram.

Each clustering is based on a different agglomeration method: the first uses the *average* method and the second uses the *complete* method. In line 11, the tanglegram is plotted. Figure 5.9 illustrates what the resulting tanglegram looks like. Note that the lines are colored to indicate that two corresponding sub-trees are included in both dendrograms. Moreover, nodes that have no corresponding node in the other dendrogram are plotted with dashed lines.

Baker [Baker (1974)] has presented a gamma index for comparing two hierarchical clusterings, which is derived from Goodman and Kruskal rank correlation measure. The index is defined as the rank correlation between the stages at which pairs of data points join in each of the two hierarchies. The index value ranges between -1 and 1, where a near-to-zero value indicates that the two hierarchies are not statistically similar. The Baker's Gamma Index can be calculated using the *cor bakers gamma* function as illustrated in List 5.13. Notice that in this example, the two dendrograms are pretty similar, and thus the Baker's Gamma Index is greater than 0.99.

The Cophenetic correlation is defined as the correlation between two cophenetic distance matrices of two dendrograms. The cophenetic distance between two data points is defined as the inter-group dissimilarity at which the two points are first joined into a single cluster. Similarly to Baker's Gamma Index, the cophenetic correlation has values between -1 and 1, where near-to-zero values indicate that the two trees are not statistically similar. The correlation can be calculated using the *cor cophenetic* function as illustrated in line 12 of List 5.13.

Listing 5.13: Using various measures for comparing a pair of dendrograms

```
1  require(graphics);
2  require(utils)
3  require(dendextend)
4
5  set.seed(1123)
6  iris2dsample <- iris[sample(nrow(iris),12),3:4]
7
8  # clustering the data
9  hc1 <- hclust(dist(iris2dsample), "ave")
10 hc2 <- hclust(dist(iris2dsample), "com")
11 cor_bakers_gamma(hc1, hc2)
12 cor_cophenetic(hc1, hc2)
```

The Fowlkes-Mallows Index [Fowlkes and Mallows (1983)] This index can get values from 0 to 1, where a higher value indicates a greater similarity between the two hierarchical clusterings. Given a hierarchical clustering, a k clusters can be produced by cutting the tree in a particular height. For each value of k, the following matrix is defined:

$$M = [m_{ij}](i = 1, \ldots, k; j = 1, \ldots, k) \tag{5.1}$$

where m_{ij} is the number of points in common between the ith cluster of the first tree and the jth cluster of the second tree. The Fowlkes-Mallows index for a particular value of k is defined as:

$$B_k = \frac{T_k}{\sqrt{P_k Q_k}} \tag{5.2}$$

where:

$$T_k = \sum_{i=1}^{k} \sum_{j=1}^{k} m_{i,j}^2 - n \tag{5.3}$$

$$P_k = \sum_{i=1}^{k} \left(\sum_{j=1}^{k} m_{i,j} \right)^2 - n \tag{5.4}$$

$$Q_k = \sum_{j=1}^{k} \left(\sum_{i=1}^{k} m_{i,j} \right)^2 - n \tag{5.5}$$

The Fowlkes-Mallows index can be calculated using the *FM Index* function provided by the *dendextend* package. Line 17 in List 5.14 compares a pair of dendrograms, both cut at k=2. The mean and variance of B_k under the assumption that the two clusterings are unrelated are also provided as *E FM* and *V FM*, respectively. If a particular B_k falls outside the limits of the confidence interval of *E FM*, then the similarity of the two clusterings is statistically significant.

In line 20, the *Bk plot* function is used to draw a scatter plot of the Fowlkes-Mallows index for a series of k cuts. It also plots the asymptotic rejection line and permutation rejection line for rejecting the null hypothesis that the two clusterings are unrelated. If the actual B_k values are located above those lines, then it can be said that the two clusterings are statistically related like in the current example (see Figure 5.10).

Listing 5.14: Using The Fowlkes-Mallows index for comparing a pair of dendrograms

```
1  require(graphics);
2  require(utils)
3  require(dendextend)
4
5  set.seed(1123)
6  iris2dsample <- iris[,-5]
7
8  # clustering the data
9  hc1 <- hclust(dist(iris2dsample), "ave")
10 hc2 <- hclust(dist(iris2dsample), "com")
11
12 # converting clusterings to dendrograms
13 dend1 <- as.dendrogram(hc1)
14 dend2 <- as.dendrogram(hc2)
15
16 # calculating FM index for a certain cut of k=2
17 FM_index(cutree(dend1, k=2), cutree(dend2, k=2))
18
19 # Ploting FM index as a function of k
20 Bk_plot(dend1, dend2, main = "Bk plot")
```

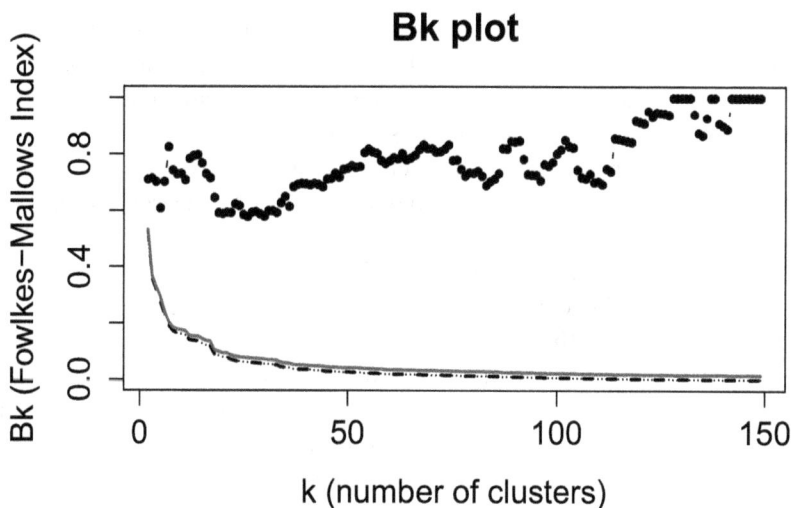

Fig. 5.10: Illustrating B_k plot.

Dendrograms have two main drawbacks [Schonlau *et al.* (2002)]. First, because each leaf is associated with only one data point, dendrogram is suitable only for small datasets. For large datasets, dendrograms become impractical. Second, dendrogram are designed for hierarchical clustering algorithms in which clusters can be successively joined, and thus, it is not always possible to use it for other clustering approaches.

5.6   Clustergram

Clustergram examines how cluster members are assigned to clusters as the number of clusters increases [Schonlau *et al.* (2002)]. In contrast to the dendrogram, this graph can be used to evaluate nonhierarchical clustering algorithms (e.g. k-means) and for hierarchical cluster algorithms when the dataset is too big to describe it using a dendrogram.

The clustergram is built as follows: The x-axis represents the number of clusters. For each number of clusters ($x = 2, \ldots, l$), a clustering structure is obtained by running a clustering algorithm such as k-means. For each cluster, an aggregate value is computed over all dimensions and over all objects in that cluster. The aggregate value can be as simple as a mean. However, other functions like principal components or weighted means could be used as well.

After calculating the aggregate values, the values of consecutive cluster structures are connected with parallelograms. The width of the parallelogram is determined by the number of instances that go from a particular cluster to the same cluster in the subsequent cluster structure.

The List 5.15 demonstrates how clustergram can be created in R using an external function code that is stored in the GitHub (the well-known Web-based Git repository hosting service). Figure 5.11 illustrates the resulting clustergram. Initially, all observations form a single cluster. This cluster is split into two clusters. The lower parallelogram is much thicker than the upper one, indicating that many more observations fall into the lower cluster. These two clusters are then split into three clusters. A new cluster is formed in the middle, which draws some observations that were previously classified in the lower cluster, and some that were previously classified in the higher cluster. Because the new cluster is formed from observations of more than one previous cluster (i.e., has more than one parent), this is a nonhierarchical split. On the vertical axis, the log base 10 of the average number of lawsuits filed against a company is shown. Therefore, "higher" or "lower" clusters refer to clusters with companies that, on average, have a larger or smaller number of lawsuits.

Listing 5.15: Using *plot* function to draw the clusters

```
1  library(RCurl)
2
3  source_https <- function(url, ...) {
4    # load package
5    # parse and evaluate each .R script
6    sapply(c(url, ...), function(u) {
7      eval(parse(text = getURL(u, followlocation = TRUE,
            cainfo = system.file("CurlSSL", "cacert.pem",
            package = "RCurl"))), envir = .GlobalEnv)
8    })
9  }
10
11 source_https("https://raw.github.com/talgalili/R-code-
        snippets/master/clustergram.r")
12
13 data(iris)
14 set.seed(250)
15 par(cex.lab = 1.5, cex.main = 1.2)
16 Data <- scale(iris[,-5]) # notice I am scaling the vectors)
17 clustergram(Data, k.range = 1:5, line.width = 0.004) #
        notice how I am using
```

Clustergram of the PCA–weighted Mean of the clusters k–mean clusters vs number of clusters (k)

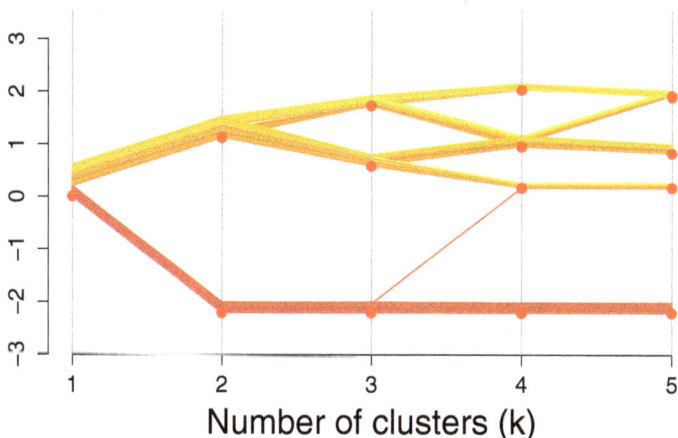

Number of clusters (k)

Fig. 5.11: Illustrating clustergram.

5.7   t-Distributed Stochastic Neighbor Embedding (t-SNE)

t-Distributed Stochastic Neighbor Embedding (t-SNE) is a dimensionality reduction technique commonly used for visualizing high-dimensional data in lower-dimensional spaces. Unlike linear techniques such as Principal Component Analysis (PCA), t-SNE aims to preserve local similarities between data points, making it particularly effective for visualizing clusters in complex datasets.

The t-SNE algorithm works by first constructing a probability distribution over pairs of high-dimensional data points, where similar points have higher probabilities of being chosen as neighbors. It then defines a similar probability distribution over the corresponding low-dimensional points in the embedding space. The goal of t-SNE is to minimize the mismatch between these two distributions using gradient descent.

t-SNE is widely used to visualize clusters in high-dimensional data. By projecting data onto a two- or three-dimensional space, t-SNE can reveal the underlying structure of the data and highlight clusters of similar points. This makes it particularly useful for exploratory data analysis and identifying patterns in complex datasets.

Consider a dataset containing images of handwritten digits. Each digit corresponds to a different class, and the goal is to visualize the distribution of digits in the dataset. By applying t-SNE to the high-dimensional image features, we can project the data onto a two-dimensional space and visualize the clusters corresponding to each digit class. This allows us to gain insights into the similarities and differences between different digits and identify any misclassifications or outliers.

The key parameters of t-SNE include:

- **Perplexity:** A hyperparameter that controls the number of nearest neighbors considered during optimization. Higher perplexity values capture global structure, while lower values focus on local structure.
- **Learning rate:** The step size of the gradient descent algorithm. A higher learning rate may lead to faster convergence but can also cause instability.
- **Number of iterations:** The number of iterations used for optimization. Increasing the number of iterations can improve the quality of the embedding but also increase computation time.

The List 5.16 demonstrates how t-SNE can be used to embed clusters in high dimensions into two dimensions. This code generates 3D sample data with two clusters, performs t-SNE dimensionality reduction using the Rtsne package, and then visualizes the resulting embedding with cluster labels using ggplot2.

Listing 5.16: Using tSNE to visualize clusters

```
1  library(Rtsne)
2  library(ggplot2)
3
4  # Generate sample data with clusters
5  set.seed(123)
6  data <- data.frame(
7    x = c(rnorm(100, mean = 0), rnorm(100, mean = 3)),
8    y = c(rnorm(100, mean = 0), rnorm(100, mean = 3)),
9    z = c(rnorm(100, mean = 0), rnorm(100, mean = 3)),
10   cluster = factor(rep(1:2, each = 100))
11 )
12
13 # Perform t-SNE
14 tsne_result <- Rtsne(as.matrix(data[, -4]), perplexity = 30)
15
16 # Visualize t-SNE embedding with clusters
17 tsne_df <- data.frame(
18   x = tsne_result$Y[, 1],
19   y = tsne_result$Y[, 2],
20   cluster = data$cluster
21 )
22
23 # Plot
24 ggplot(tsne_df, aes(x = x, y = y, color = cluster)) +
25   geom_point(size = 3) +
26   scale_color_manual(values = c("blue", "red")) +  #
          Customize cluster colors if needed
27   theme_minimal()
```

5.7.1 *Advantages and limitations*

t-SNE offers several advantages for cluster visualization:

- **Non-linearity:** t-SNE preserves the local structure and can reveal intricate patterns in high-dimensional data that linear techniques may miss.

- **Flexibility:** t-SNE can be applied to various data types, including numerical, categorical, and text data.
- **Intuitive visualization:** The low-dimensional embeddings produced by t-SNE are easy to interpret and can provide valuable insights into the underlying data structure.

However, t-SNE also has limitations:

- **Sensitivity to parameters:** t-SNE performance can be sensitive to the choice of parameters, such as perplexity and learning rate, which may require careful tuning.
- **Computationally intensive:** t-SNE can be computationally expensive, especially for large or high-dimensional data.
- **Noisy embeddings:** In some cases, t-SNE may produce noisy embeddings or distort the underlying data structure, particularly for datasets with complex or overlapping clusters.

Chapter 6

Cluster Validity: Evaluation of Clustering Algorithms

6.1 Introduction

Assessing the quality and effectiveness of a clustering algorithm's results is known as cluster validity evaluation. Determining whether a particular clustering outcome is good or not is a contentious issue. In fact, as early as 1964, Bonner argued that there is no universal definition for what constitutes a good clustering [Bonner (1964)]. The evaluation of clustering results remains largely subjective and dependent on the evaluator's perspective. While no universally accepted definition exists for what constitutes a good clustering, researchers have developed various approaches to evaluate clustering results. These approaches can be broadly classified as internal, external, or relative, each with its own strengths and limitations, and the choice depends on the specific requirements and constraints of the problem at hand.

Internal evaluation methods assess the clusters based on quantitative measures derived from the data. External evaluation methods involve comparing the clusters to a predefined cluster structure, which serves as a ground truth or gold standard. Relative evaluation methods compare the evaluated cluster schema to other clustering schemes, typically obtained using the same algorithm but with different hyperparameter settings.

6.2 Internal Criteria

Internal quality metrics usually measure the compactness of the clusters using some similarity measure. It usually measures the intra-cluster homogeneity, the inter-cluster separability or a combination of these two. It does not use any external information besides the data itself.

6.2.1 *Sum of squared error (SSE)*

Let $\mu^{\{k\}}$ denotes the barycenter of the points in the cluster C_k and by μ the barycenter of all the points in the dataset. $\mu^{\{k\}}$ and μ are row-vectors with length n:

$$\mu^{\{k\}} = \frac{1}{N_k} \sum_{\mathbf{x_i} \in C_k} \mathbf{x_i}$$

$$\mu = \frac{1}{m} \sum_{i=1}^{m} \mathbf{x_i}$$

where N_k is the number of instances belonging to cluster k. Note that:

$$\sum_{k=1}^{K} N_k = m$$

where m is the total number of data instances.

SSE is the simplest and most widely used criterion measure for clustering. It is calculated as:

$$SSE = \sum_{k=1}^{K} \sum_{i=1}^{N_k} \|\mathbf{x_i} - \mu_k\|^2$$

Clustering methods that minimize the SSE criterion are often called minimum variance partitions, since by simple algebraic manipulation the SSE criterion may be written as:

$$SSE = \frac{1}{2} \sum_{k=1}^{K} N_k \bar{S}_k$$

where:

$$\bar{S}_k = \frac{1}{N_k^2} \sum_{x_i, x_j \in C_k} \|x_i - x_j\|^2$$

(C_k = cluster k).

The SSE criterion function is suitable for cases in which the clusters form compact clouds that are well separated from one another [Duda *et al.* (2012)].

6.2.2 The Ball-Hall index

The mean dispersion of a cluster is the mean of the squared distances of the points of the cluster with respect to their barycenter. The Ball-Hall index is the mean, through all the clusters, of their mean dispersion:

$$BHI = \frac{1}{K} \sum_{k=1}^{K} \frac{1}{N_k} \sum_{i=1}^{N_k} \| \mathbf{x_i} - \mu_k \|^2$$

6.2.3 Other minimum variance criteria

Additional minimum criteria for SSE may be produced by replacing the value of S_k with expressions such as:

$$\bar{S}_k = \frac{1}{N_k^2} \sum_{x_i,x_j \in C_k} s(x_i, x_j)$$

Or:

$$\bar{S}_k = \min_{x_i,x_j \in C_k} s(x_i, x_j)$$

6.2.4 Scatter criteria

The scalar scatter criteria are derived from the scatter matrices, reflecting the within-cluster scatter, the between-cluster scatter and their summation — the total scatter matrix. For the kth cluster, the scatter matrix is a square symmetric matrices defined as:

$$S_k = \sum_{x \in C_k} (x - \mu_k)(x - \mu_k)^T$$

The within-cluster scatter matrix is calculated as the summation of the last definition over all clusters:

$$S_W = \sum_{k=1}^{K} S_k$$

The between-cluster scatter matrix may be calculated as:

$$S_B = \sum_{k=1}^{K} N_k (\mu_k - \mu)(\mu_k - \mu)^T$$

where μ is the total mean vector and is defined as:

$$\mu = \frac{1}{m} \sum_{k=1}^{K} N_k \mu_k$$

The total scatter matrix should be calculated as follows:

$$S_T = \sum_{x \in C_1, C_2, \ldots, C_K} (x - \mu)(x - \mu)^T$$

Three scalar criteria may be derived from S_W, S_B and S_T:

- **The trace criterion** — The sum of the diagonal elements of a matrix. Minimizing the trace of S_W is similar to minimizing SSE and is therefore acceptable. This criterion, representing the within-cluster scatter, is calculated as:

$$J_e = tr[S_W] = \sum_{k=1}^{K} \sum_{x \in C_k} \|x - \mu_k\|^2$$

Another criterion, which may be maximized, is the between cluster criterion:

$$tr[S_B] = \sum_{k=1}^{K} N_k \|\mu_k - \mu\|^2$$

- **The determinant criterion** — The determinant of a scatter matrix roughly measures the square of the scattering volume. Since SB will be singular if the number of clusters is less than or equal to the dimensionality, or if $m - c$ is less than the dimensionality, its determinant is not an appropriate criterion. If we assume that SW is nonsingular, the determinant criterion function using this matrix may be employed:

$$J_d = \det(S_W) = |S_W| = \left| \sum_{k=1}^{K} S_k \right|$$

- **The invariant criterion** — The eigenvalues $\lambda_1, \lambda_2, \ldots, \lambda_d$ of

$$S_W^{-1} S_B$$

are the basic linear invariants of the scatter matrices. Good partitions are ones for which the nonzero eigenvalues are large. As a result, several criteria may be derived, including the eigenvalues. Three such criteria are:

(1) $tr[S_W^{-1} S_B] = \sum_{i=1}^{d} \lambda_i$

(2) $J_f = tr[S_T^{-1} S_W] = \sum_{i=1}^{d} \frac{1}{1+\lambda_i}$

(3) $\frac{|S_W|}{|S_T|} = \prod_{i=1}^{d} \frac{1}{1+\lambda_i}$

6.2.5   C *index*

The C index is defined as:

$$C = \frac{s_W - s_{\min}}{s_{\max} - s_{\min}}$$

where s_W is the sum of the distances between all the pairs of points inside each cluster:

$$s_W = \sum_{k=1}^{K} \sum_{\mathbf{x_i},\mathbf{x_i} \in C_k} d(\mathbf{x_i},\ \mathbf{x_j})$$

and $s_{\max}$ is the sum of the N_W largest distances between all the pairs of points in the entire dataset, and $s_{\min}$ is the sum of the N_W smallest distances between all the pairs of points in the entire dataset. Note that in the cluster C_k, there are $N_k(N_k - 1)/2$ pairs of distinct points and N_W is defined as:

$$N_W = \sum_{k=1}^{K} \frac{N_k(N_k - 1)}{2}$$

While in total, there are N_T distances between all pairs of points in the dataset:

$$N_T = \frac{m(m-1)}{2}$$

where $s_{\max}$ and $s_{\min}$ takes into consideration only N_W of the N_T pairs.

6.2.6   *The McClain-Rao index*

The McClain-Rao index is defined as the ratio between the mean within-cluster and between-cluster distances:

$$MR = \frac{N_B}{N_W} \frac{s_W}{s_B}$$

where s_W and N_W were defined in the previous section, $N_B = m(m-1)/2 - N_W$ represents the total number of distances between pairs of points which do not belong to the same cluster and s_B is the sum of the between-cluster distances:

$$s_B = \sum_{k<k'} \sum_{\mathbf{x_i} \in C_k,\mathbf{x_j} \in C_{k'}} d(\mathbf{x_i},\ \mathbf{x_j})$$

6.2.7 The Banfeld-Raftery index

This index is the weighted sum of the logarithms of the traces of the variance-covariance matrix of each cluster: The index can be written like this:

$$BRI = \sum_{k=1}^{K} n_k \log\left(\frac{\text{Tr}(S_k)}{N_k}\right)$$

The quantity $\text{Tr}(S_k)/N_k$ is the mean of the squared distances between the observations in cluster C_k and their barycenter μ_k. If a cluster contains a single observation, this trace equals 0, and the logarithm is undefined.

6.2.8 Condorcet's criterion

Another appropriate approach is to use the Condorcet's solution (1785) to the ranking problem [Marcotorchino and Michaud (1979)]. In this case, the criterion is calculated as follows:

$$\sum_{C_i \in C} \sum_{\substack{x_j, x_k \in C_i \\ x_j \neq x_k}} s(x_j, x_k) + \sum_{C_i \in C} \sum_{x_j \in C_i; x_k \notin C_i} d(x_j, x_k)$$

where $s(x_j, x_k)$ and $d(x_j, x_k)$ measures the similarity and distance of the vectors x_j and x_k respectively.

6.2.9 The C-criterion

The C-criterion [Fortier and Solomon (1966)] is an extension of the Condorcet's criterion and is defined as:

$$\sum_{C_i \in C} \sum_{\substack{x_j, x_k \in C_i \\ x_j \neq x_k}} (s(x_j, x_k) - \gamma) + \sum_{C_i \in C} \sum_{x_j \in C_i; x_k \notin C_i} (\gamma - s(x_j, x_k))$$

where γ is a threshold value.

6.2.10 The Calinski-Harabasz index

Using the above notations, the Calinski-Harabasz index is defined as:

$$C = \frac{\text{Tr}(S_B)/(K-1)}{\text{Tr}(S_W)/(m-K)} = \frac{m-K}{K-1}\frac{\text{Tr}(S_B)}{\text{Tr}(S_W)}$$

where S_B is the between-group scatter matrix and S_W is within-group scatter matrix and $\text{Tr}(S_B)$ and $\text{Tr}(S_W)$ are their traces respectively.

6.2.11   *The Silhouette index*

The value $a(i)$ denotes the within-cluster mean distance of point $\mathbf{x_i}$ to all other points of the cluster it belongs to:

$$a(i) = \frac{1}{N_k - 1} \sum_{\mathbf{x_j}, \mathbf{x_i} in C_k; \mathbf{x_j} \neq \mathbf{x_i}} d(\mathbf{x_i}, \mathbf{x_j})$$

Let $\beta(\mathbf{x_i}, C_{k'})$ represents the distance of the point $\mathbf{x_i}$ to the points in another cluster $C_{k'}$:

$$\beta(\mathbf{x_i}, C_{k'}) = \frac{1}{N_{k'}} \sum_{\mathbf{x_j} \in C_{k'}} d(\mathbf{x_i}, \mathbf{x_i})$$

The value $b(i)$ denotes the smallest of these mean distances:

$$b(i) = \min_{\forall k' \neq k | \mathbf{x_i} \in C_k} \beta(\mathbf{x_i}, C_{k'})$$

Note that C_k is the cluster that $\mathbf{x_i}$ belongs to, and k' represents any cluster other than C_k.

The Silhouette width of the point $\mathbf{x_i}$ is defined as the following ratio:

$$s(i) = \frac{b(i) - a(i)}{\max(a(i), b(i))}$$

Note that $s(i)$ can get a value between -1 and 1. More specifically, a value of 1 indicates that the point $\mathbf{x_i}$ is assigned to the correct cluster whereas a value of -1 indicates that the $\mathbf{x_i}$ is assigned to an incorrect cluster and, thus, should be reassigned to another cluster.

The mean of the Silhouette widths for all points in a particular cluster C_k is called the cluster mean Silhouette and is defined as:

$$\mathfrak{s}_k = \frac{1}{N_k} \sum_{\mathbf{x_i} \in C_k} s(i)$$

The Silhouette index is defined as the mean of all clusters Silhouette widths:

$$C = \frac{1}{K} \sum_{k=1}^{K} \mathfrak{s}_k$$

6.2.12   *Log SS ratio index*

Similarly to the Calinski-Harabasz index, the Log SS Ratio index also uses the traces of the between-group scatter matrix and the within-group scatter matrix. Formally, it is defined as:

$$LOGSSRatio = \log\left(\frac{\mathrm{Tr}(S_B)}{\mathrm{Tr}(S_W)}\right)$$

6.2.13   *The Dunn index*

Let $d_{\max}$ denotes the largest within-cluster distance. Moreover, let $d_{\min}$ denotes the minimal distance between observations of different clusters.

For each cluster C_k, the value D_k is the diameter of the cluster, i.e. the largest distance separating two observations in the cluster:

$$D_k = \max_{\mathbf{x_i},\mathbf{x_j}\in C_k} ||\mathbf{x_i} - \mathbf{x_j}||$$

Then $d_{\max}$ is the largest of these distances D_k:

$$d_{\max} = \max_{1\le k\le K} D_k$$

The distance between clusters C_k and $C_{k'}$ is measured by the distance between their closest points:

$$d_{kk'} = \min_{x_i\in C_k; x_j\in C_{k'}} ||x_i - x_j||$$

and $d_{\min}$ is the smallest of these distances $d_{kk'}$:

$$d_{\min} = \min_{k\ne k} d_{kk'}$$

Based on the above values, we can define the Dunn index as the ratio of $d_{\min}$ and $d_{\max}$, i.e.:

$$DI = \frac{d_{\min}}{d_{\max}}$$

A higher Dunn index indicates better clustering.

6.2.14   *The generalized Dunn index (GDI)*

The generalized Dunn index (or GDI for short) uses different measures to evaluate the distances between clusters and within groups. The GDI index is defined as:

$$GDI = \frac{\min_{k \neq k'} \delta(C_k, C_{k'})}{\max_k \triangle(C_k)}$$

where δ denotes a measure for between-cluster distance and $\triangle$ denotes a measure for within-cluster distance.

Three different definitions of $\triangle$ and six different definitions of δ have been presented in the literature. Together we get in total 18 different generalized Dunn indices.

The definitions of the within-cluster distances $\triangle$ are:

$$\triangle_1(C_k) = \max_{\mathbf{x_i}, \mathbf{x_j} \in C_k, i \neq j} d(\mathbf{x_i}, \mathbf{x_j})$$

$$\triangle_2(C_k) = \frac{1}{N_k(N_k - 1)} \sum_{\mathbf{x_i}, \mathbf{x_j} \in C_k, i \neq j} d(\mathbf{x_i}, \mathbf{x_j})$$

$$\triangle_3(C_k) = \frac{2}{N_k} \sum_{\mathbf{x_i} \in C_k} d(\mathbf{x_i}, \mu_k)$$

The definitions of the between-cluster distances δ are:

$$\delta_1(C_k, C_{k'}) = \min_{\mathbf{x_i} \in C_k, \mathbf{x_j} \in C_{k'}} d(\mathbf{x_i}, \mathbf{x_j})$$

$$\delta_2(C_k, C_{k'}) = \max_{\mathbf{x_i} \in C_k, \mathbf{x_j} \in C_{k'}} d(\mathbf{x_i}, \mathbf{x_j})$$

$$\delta_3(C_k, C_{k'}) = \frac{1}{N_k N_{k'}} \sum_{\mathbf{x_i} \in C_k, \mathbf{x_j} \in C_{k'}} d(\mathbf{x_i}, \mathbf{x_j})$$

$$\delta_4(C_k, C_{k'}) = d(\mu_k, \mu_{k'})$$

$$\delta_5(C_k, C_{k'}) = \frac{1}{N_k + N_{k'}} \left(\sum_{\mathbf{x_i} \in C_k} d(\mathbf{x_i}, \mu_k) + \sum_{\mathbf{x_j} \in C_{k'}} d(\mathbf{x_j}, \mu_{k'}) \right)$$

$$\delta_6(C_k, C_{k'}) = \max \left\{ \sup_{\mathbf{x_i} \in C_k} \inf_{\mathbf{x_j} \in C_{k'}} d(\mathbf{x_i}, \mathbf{x_j}), \sup_{\mathbf{x_j} \in C_{k'}} \inf_{\mathbf{x_i} \in C_k} d(\mathbf{x_i}, \mathbf{x_j}) \right\}$$

Note that the measures δ_1 to δ_4 have already been referred in previous chapters as single linkage, complete linkage, average linkage, and centroid linkage, respectively. In measure δ_5 we weigh the mean distances between the observations in clusters C_k and $C_{k'}$ and their corresponding centers. The measure δ_6 is known in the literature as the Hausdorff distance or the Pompeiu–Hausdorff distance and it measures the greatest of all the distances from a point in one cluster to the closest point in the other cluster.

6.2.15 The Davies-Bouldin index

Let δ_k denotes the mean distance of the observations belonging to cluster C_k to their center ν_k:

$$\delta_k = \frac{1}{N_k} \sum_{\mathbf{x_i} \in C_k} ||\mathbf{x_i} - \mu_k||$$

Let:

$$\triangle_{kk'} = d(\mu_k, \, \mu_{k'}) = ||\mu_k - \mu_{k'}||$$

denotes the distance between the centers μ_k and $\mu_{k'}$ of clusters C_k and $C_{k'}$. For each pair of clusters k and k' we can calculate the ratio:

$$\frac{\delta_k + \delta_{k'}}{\triangle_{kk}}$$

For each cluster, we can calculate the maximum ratio over all other clusters for which $k' \neq k$. The Davies-Bouldin index is defined as the mean value of these maximum ratios:

$$DB = \frac{1}{K} \sum_{k=1}^{K} \max_{k} \left(\frac{\delta_k + \delta_{k'}}{\triangle_{kk}} \right)$$

A lower value of the Davies-Bouldin index means that the clustering is better. Moreover, due to the way it is defined, the Davies-Bouldin index is symmetric and non-negative.

6.2.16   *The Baker-Hubert Gamma index*

In cluster validity, the Γ index can have a value between -1 and 1 and is defined as:

$$\Gamma = \frac{s^+ - s^-}{s^+ + s^-}$$

where s^+ represents the number of times that a pair of points not assigned to the same cluster had a larger separation than a pair that were assigned to the same cluster. The notation s^- represents the reverse outcome, i.e. the distance between two points that are assigned to the same cluster is greater than a distance between two points not belonging to the same cluster.

6.2.17   *The G-plus index*

The G-plus index is defined as the number of discordant pairs divided by all distinct point pairs. Using the notation presented in the last section, the G-plus index is defined as:

$$G_+ = \frac{2s^-}{N_T(N_T - 1)}$$

6.2.18   *The Det-ratio index*

The Det-ratio index is defined as:

$$C = \frac{\det(S_T)}{\det(S_W)}$$

where S_T and S_W are the total scatter and within-cluster scatter matrices correspondingly.

6.2.19   *The log Det ratio index*

The log Det ratio index is a logarithmic variant of the Det-ratio index presented in the last section. Formally, it is defined as:

$$LDR = m \log \left(\frac{\det(S_T)}{\det(S_W)} \right)$$

where S_W is the within-cluster scatter matrix and S_T is the scatter matrix.

6.2.20   The $k^2|W|)$ index

The $k^2|W|)$ index is defined as:

$$C = K^2 \det(S_W)$$

where S_W is the within-cluster scatter matrix.

6.2.21   Category utility metric

The category utility [Gluck (1985)] is defined as the increase of the expected number of feature values that can be correctly predicted given a certain clustering. This metric is useful for problems that contain a relatively small number of nominal features, each having a small cardinality.

6.2.22   Edge cut metrics

In some cases, it can be useful to represent the clustering problem as an edge cut minimization problem. In this case, the quality is measured by the ratio of the remaining edge weights to the total precut edge weights. If there is no restriction on the size of the clusters, finding the optimal value is easy. As a result, the min-cut measure has been revised to penalize imbalanced structures.

6.3   External Quality Criteria

External measures can be useful to examine whether the structure of the clusters match some predefined classification of the instances.

6.3.1   Mutual information based measure

The mutual information criterion can be used as an external measure for clustering [Strehl *et al.* (2000)]. The measure for m instances clustered using $\mathbf{C} = \{C_1, \ldots, C_g\}$ and referring to the target attribute y whose domain is $dom(y) = \{c_1, \ldots, c_k\}$ is defined as follows:

$$C = \frac{2}{m} \sum_{l=1}^{g} \sum_{h=1}^{k} m_{l,h} \log_{g \cdot k} \left(\frac{m_{l,h} \cdot m}{m_{.,l} \cdot m_{l,.}} \right)$$

where $m_{l,h}$ indicates the number of instances that are in cluster C_l and also in class c_h. $m_{.,h}$ denotes the total number of instances in the class c_h. Similarly $m_{l,.}$ indicates the number of instances in cluster C_l.

6.3.2   *Precision-recall measure*

Precision and recall, commonly used metrics in classification tasks, can also be adapted for external validation of clustering results. Here is how they can be applied:

For clustering external validation, we compare the clustering results with ground truth (predefined labels). The metrics assess how well the clustering algorithm's output aligns with these predefined labels.

In the context of clustering, precision measures the accuracy of the clustering assignments. It is the proportion of data points that are correctly assigned to a cluster among all the data points assigned to that cluster. Recall measures the completeness of the clustering assignments. It is the proportion of data points that are correctly assigned to a cluster among all the data points that should have been assigned to that cluster.

Calculating precision and recall for clustering

Pairwise precision and recall

- **True positive (TP)**: A pair of data points that are in the same cluster in the predicted clustering and in the same cluster in the ground truth.
- **False positive (FP)**: A pair of data points that are in the same cluster in the predicted clustering but in different clusters in the ground truth.
- **False negative (FN)**: A pair of data points that are in different clusters in the predicted clustering but in the same cluster in the ground truth.

Based on these definitions, pairwise precision and recall can be calculated as follows:

$$\text{Precision} = \frac{TP}{TP + FP}$$

$$\text{Recall} = \frac{TP}{TP + FN}$$

Cluster-level precision and recall

- For each cluster in the predicted clustering, determine the dominant ground truth label for the cluster.
- Calculate precision and recall for each cluster by comparing the predicted cluster with the corresponding ground truth cluster.

Adjusted mutual information (AMI) and F-measure

To further assess clustering performance:

- **Adjusted mutual information (AMI):** Adjusts the Mutual Information score to account for chance. It ranges from -1 (no agreement) to 1 (perfect agreement).
- **F-measure (F1 Score):** [Larsen and Aone (1999)] The harmonic mean of precision and recall, calculated as:

$$F1 = 2 \times \frac{\text{Precision} \times \text{Recall}}{\text{Precision} + \text{Recall}}$$

6.3.3   *Rand index*

The Rand index [Rand (1971)] is a simple criterion used to compare an induced clustering structure C_1 to a given clustering structure C_2. Let a be the number of pairs of instances that are assigned to the same cluster in C_1 and in the same cluster in C_2, b be the number of pairs of instances that are in the same cluster in C_1, but not in the same cluster in C_2, c be the number of pairs of instances that are in the same cluster in C_2, but not in the same cluster in C_1, and d be the number of pairs of objects that are assigned to different clusters in C_1 and C_2. The variables a and d can be interpreted as agreements, and b and c as disagreements. Thus, the Rand index is defined as:

$$RAND = \frac{a+d}{a+b+c+d} \tag{6.1}$$

The Rand index gets values between 0 and 1. When the two clusterings agree perfectly, the Rand index is 1.

A problem with the Rand index is that its expected value of two random clusterings do not take a constant value (such as zero). The adjusted Rand index aims to overcome this disadvantage [Hubert and Arabie (1985)]. Thus, it gets a value close to 0.0 for two random clusterings and exactly 1.0 when the clusterings C_1 and C_2 are identical (up to a permutation) independently of the number of clusters and samples. Formally, the adjusted Rand index is defined as:

$$ARI = \frac{RAND - EXPECTED\_INDEX}{MAX\_INDEX - EXPECTED\_INDEX}$$

where $RAND$ refers to the unadjusted Rand index provided in Eq. (6.1). More specifically, assuming a maximum Rand index of 1.0, the adjusted Rand index becomes:

$$\frac{\sum_{i,j}\binom{n_{ij}}{2} - \sum_{i}\binom{n_{i.}}{2}\sum_{j}\binom{n_{.j}}{2}\Big/\binom{n}{2}}{\frac{1}{2}\left[\sum_{i}\binom{n_{i.}}{2} + \sum_{j}\binom{n_{.j}}{2}\right] - \sum_{i}\binom{n_{i.}}{2}\sum_{j}\binom{n_{.j}}{2}\Big/\binom{n}{2}} \qquad (6.2)$$

where n is the number of instances, $n_{i,j}$ denotes the number of instances in common between cluster i of the first clustering $\mathbf{C_1}$ and cluster j of the second clustering $\mathbf{C_2}$. Moreover according the notation convention, $n_{i,.}$ and $n_{.,j}$ denote the total number of instances in cluster i of $\mathbf{C_1}$ and cluster j of $\mathbf{C_2}$ respectively.

It should be noted that the adjusted Rand index can yield negative values if the index is less than the expected index.

6.3.4 Folkes and Mallows index

Folkes and Mallows index is defined as:

$$FM = a/\sqrt{m_1 \cdot m_2} = \sqrt{\frac{a}{a+b} \cdot \frac{a}{a+c}} \qquad (6.3)$$

where $m_1 = (a + b)$, $m_2 = (a + c)$.

The higher the value of the index is the more similar clustering structure $\mathbf{C_1}$ and $\mathbf{C_2}$ are.

6.4 Calculating Validity Indices in R

The *clusterCrit* package in R provides the implementation of many of the internal and external indices described in the previous sub-sections. By calling the function *getCriteriaNames(TRUE)*, we will get a list that specifies all supported internal indices. Similarly calling *getCriteriaNames(FALSE)* returns all supported external indices. The code 6.1 illustrates the usage of an external index for evaluating a clustering. In line 3, we create a sample from the Iris dataset. In line 4, we create a first clustering using k-means algorithm assuming three centroids and an Euclidean distance measure. Then, in line 5, we estimate the Rand index by comparing the resulting clustering with the original labeling, which can be found in the fifth column (Species) of the Iris dataset. Note that the function *as.integer* is used to convert the string Species' labels into integers, before it can be compared to the resulted clustering. In lines 7-8, we show how external index can be used for comparing two different clusterings (*c1* and *c2* in this case).

Listing 6.1: Using external index for comparing two clustering results

```
1 library(amap)
2 library(clusterCrit)
3 iris2dsample <- iris[sample(nrow(iris),40),]
4 c1 <- Kmeans(iris2dsample[,1:4], centers=3, method="
    euclidean",iter.max = 100)
5 Rand_Index_1=extCriteria(c1$cluster,as.integer(iris2dsample
    [,5]), crit="Rand")
6
7 c2 <- Kmeans(iris2dsample[,1:4], centers=4, method="maximum"
    ,iter.max = 10)
8 Rand_Index_2=extCriteria(c1$cluster, c2$cluster, crit="Rand"
    )
```

The code 6.2 illustrates the usage of an internal index for evaluating a clustering. In line 5, the Dunn index is estimated. Note that the *intCriteria* gets both the original dataset and the resulting clustering. In line 6 all supported indices are calculated by providing the value *'all'* for the *crit* parameter.

Listing 6.2: Using internal index for evaluating a clustering

```
1 library(amap)
2 library(clusterCrit)
3 iris2dsample <- iris[sample(nrow(iris),40),]
4 c1 <- Kmeans(iris2dsample[,1:4], centers=3, method="
    euclidean",iter.max = 100)
5 intCriteria(as.matrix(iris2dsample[,1:4]),c1$cluster,"Dunn")
6 intCriteria(as.matrix(iris2dsample[,1:4]),c1$cluster,"all")
```

6.5   Determining the Number of Clusters

Many clustering algorithms require that the number of clusters will be preset by the user. It is well known that this parameter affects the performance of the algorithm significantly. This poses a serious question as to which K should be chosen when prior knowledge regarding the cluster quantity is unavailable.

Note that most of the criteria that have been used to lead the construction of the clusters (such as SSE) are monotonically decreasing in K. Therefore, using these criteria for determining the number of clusters results in a trivial clustering in which each instance is a cluster by itself. Consequently, different criteria are needed to be applied here. Many methods were offered to determine which K is preferable. These methods are usually heuristics, involving the calculation of clustering criteria measures for different values of K, thus enabling the evaluation of which K was preferable.

6.5.1 *Methods based on intra cluster scatter*

Many of the methods for determining K are based on the intra-cluster (within-cluster) scatter. This category includes the within cluster depression decay [Tibshirani *et al.* (2001)], which computes an error measure W_K, for each K chosen, as follows:

$$W_K = \sum_{k=1}^{K} \frac{1}{2N_k} D_k$$

where D_k is the sum of pairwise distances for all instances in cluster k:

$$D_k = \sum_{x_i, x_j \in Ck} \|x_i - x_j\|$$

In general, as the number of clusters increases, the within cluster decay first declines rapidly. From a certain K the curve flattens. That elbow is considered to be the appropriate K according to this method.

Other heuristics relate to the intra-cluster distance as the sum of squared Euclidean distances between the data points and their cluster centres (the sum of squares error the algorithm attempts to minimize). They range from simple methods, such as the PRE method, to more sophisticated, statistics-based methods.

An example of a simple method which works well in most databases is, as mentioned above, the proportional reduction in error (PRE) method. PRE is the ratio of reduction in the sum of squares to the previous sum of squares when comparing the results of using $K + 1$ clusters to the results of using K clusters. Increasing the number of clusters by 1 is justified for PRE rates of about 0.4 or larger.

It is also possible to examine the SSE decay, which behaves similarly to the within cluster depression described above. The manner of determining K according to both measures is also similar.

An approximate F statistic can be used to test the significance of the reduction in the sum of squares as we increase the number of clusters [Hartigan (1975)]. The method obtains this F statistic as follows:

Suppose that $P(m, k)$ is the partition of m instances into k clusters, and $P(m, k+1)$ is obtained from $P(m, k)$ by splitting one of the clusters. Also, assume that the clusters are selected without regard to $x_{qi} \sim N(\mu_i, \sigma^2)$ independently over all q and i. Then, the overall mean square ratio is calculated and distributed as follows:

$$R = \left(\frac{e(P(m, k)}{e(P(m, k + 1)} - 1 \right) (m - k - 1) \approx F_{N, N(m-k-1)}$$

where $e(P(m, k))$ is the sum of squared Euclidean distances between the data instances and their cluster centers.

In fact, this F distribution is not accurate since it is based on inaccurate assumptions:

- K-means is not a hierarchical clustering algorithm, but a relocation method. Therefore, the partition $P(m, k+1)$ is not necessarily obtained by splitting one of the clusters in $P(m, k)$.
- Each x_{qi} influences the partition.
- The assumptions as to the normal distribution and independence of x_{qi} are not valid in all databases.

As the F statistic, as described above, is imprecise, Hartigan offers a crude rule of thumb: only large values of the ratio (say, larger than 10) justify increasing the number of partitions from K to $K + 1$.

6.5.2   Methods based on both the inter and intra cluster scatter

All the methods described so far for estimating the number of clusters are quite reasonable. However, they all suffer the same deficiency: None of these methods examines the inter-cluster distances. Thus, if the K-means algorithm partitions an existing distinct cluster in the data into sub-clusters (which is undesired), it is possible that none of the above methods would indicate this situation.

In light of this observation, it may be preferable to minimize the intra-cluster scatter and at the same time maximize the inter-cluster scatter. One way to achieve this goal is to define a measure that equals the ratio of intra-cluster scatter and inter-cluster scatter [Ray and Turi (1999)]. Minimizing this measure is equivalent to both minimizing the intra-cluster scatter and maximizing the inter-cluster scatter.

Another method for evaluating the "optimal" K using both inter and intra cluster scatter is the validity index method that examines two measures [Do-Jong *et al.* (2001)]:

- MICD — Mean intra-cluster distance; defined for the kth cluster as:

$$MD_k = \sum_{x_i \in C_k} \frac{\|x_i - \mu_k\|}{N_k}$$

- ICMD — Inter-cluster minimum distance; defined as:

$$d_{\min} = \min_{i \neq j} \|\mu_i - \mu_j\|$$

in order to create a cluster validity index, the behavior of these two measures around the actual number of clusters (K^*) should be used.

When the data are under-partitioned $(K < K^*)$, at least one cluster maintains large MICD. As the partition state moves towards over-partitioned $(K > K^*)$, the large MICD abruptly decreases.

The ICMD is large when the data are under-partitioned or optimally partitioned. It becomes very small when the data enters the over-partitioned state, since at least one of the compact clusters is subdivided.

Two additional measure functions may be defined in order to find the under-partitioned and over-partitioned states. These functions depend, among other variables, on the vector of the clusters centers $\mu = [\mu_1, \mu_2, \ldots \mu_K]^T$:

(1) Under-partition measure function:

$$v_u(K, \mu; X) = \frac{\sum_{k=1}^{K} MD_k}{K} \quad 2 \leq K \leq K_{\max}$$

This function has very small values for $K \geq K^*$ and relatively large values for $K < K^*$. Thus, it helps determine whether the data is under-partitioned.

(2) Over-partition measure function:

$$v_o(K, \mu) = \frac{K}{d_{\min}} \quad 2 \leq K \leq K_{\max}$$

This function has very large values for $K \geq K^*$, and relatively small values for $K < K^*$. Thus, it helps determine whether the data is over-partitioned.

The validity index uses the fact that both functions have small values only at $K = K^*$. The vectors of both partition functions are defined as follows:

$$V_u = [v_u(2, \mu; X), \ldots, v_u(K_{\max}, \mu; X)]$$

$$V_o = [v_o(2, \mu), \ldots, v_o(K_{\max}, \mu)]$$

Before finding the validity index, each element in each vector is normalized to the range $[0,1]$, according to its minimum and maximum values. For instance, for the V_u vector:

$$v_u^*(K, \mu; X) = \frac{v_u(K, \mu; X)}{\max\limits_{K=2,\ldots,K_{\max}} \{v_u(K, \mu; X)\} - \min\limits_{K=2,\ldots,K_{\max}} \{v_u(K, \mu; X)\}}$$

The process of normalization is done the same way for the V_o vector. The validity index vector is calculated as the sum of the two normalized vectors:

$$v_{sv}(K, \mu; X) = v_u^*(K, \mu; X) + v_o^*(K, \mu)$$

Since both partition measure functions have small values only at $K = K^*$, the smallest value of v_{sv} is chosen as the optimal number of clusters.

6.5.3 Criteria based on probabilistic methods

When clustering is performed using a density-based method, the determination of the most suitable number of clusters K becomes a more tractable task as a clear probabilistic foundation can be used. The question is whether adding new parameters results in a better way of fitting the data by the model. In Bayesian theory, the likelihood of a model is also affected by the number of parameters which are proportional to K. Suitable criteria that can used here include BIC (Bayesian Information Criterion), MML (Minimum Message Length) and MDL (Minimum Description Length).

6.6 Hypothesis Testing in Cluster Validity

Cluster validity aims to test if the objects in a given data set are randomly distributed in the space or if there is a particular structure according to which the objects are distributed. Like any other statistical test, we first define the Null Hypothesis, denoted as H_O. In this case, H_0 states that the objects are randomly structured. H_0 needs to be rejected in order to show that the objects form a structure.

To test the hypothesis, we use Monte Carlo techniques, which aim to estimate the probability density function (*pdf*) of a given validity index using a simulation. The process begins by calculating the validity index for our dataset, denoted as q.

We proceed by syntheticly generating many data sets, denoted as X_i, according to a normal distribution. For each data set we calculate the designated validity index, denoted as q_i. Then, based on the values of q_i, we can approximate the probability density function of this index. The function can have three possible shapes: two-tailed, left-tailed and right-tailed. Depending on the shape, we define a critical interval $\bar{D}_\alpha$, corresponding to significant level of α.

The probability density function of a statistic index q, under Ho, has a single maximum, and the region is either a half line, or a union of two half lines. Assuming that the scatter-plot has been generated using r-values of the index q, called q_i, in order to accept or reject the *Null Hypothesis Ho*, we use the procedure presented in Figure 6.1.

1: **if** the shape is right-tailed **then**
2:   **if** q's value of our data set, is greater than $(1 - \rho) \cdot r$ of q_i values **then**
3:     Reject Ho
4:   **else**
5:     Accept Ho
6:   **end if**
7: **else if** the shape is left-tailed **then**
8:   **if** q's value for our data set, is smaller than $\rho \cdot r$ r of q_i values **then**
9:     Reject Ho
10:   **else**
11:     Accept Ho
12:   **end if**
13: **else if** the shape is two-tailed **then**
14:   **if** q is greater than $(\rho/2) \cdot r$ number of q_i values and smaller than $(1 - \rho/2) \cdot r$ of q_i values **then**
15:     Accept Ho
16:   **end if**
17: **end if**

Fig. 6.1: Procedure for hypothesis testing in cluster validity.

Chapter 7

Mixture Densities-Based Clustering

7.1 Introduction

In this chapter, we will discuss two related types of clustering algorithms: The density-based clustering methods (such as DBSCAN) where clusters are detected as areas of high density separated by sparse areas and model-based clustering algorithms where the data are assumed to come from a mixture of probability distributions, each of which represents a different cluster.

7.2 DBSCAN Algorithm

DBSCAN (density-based spatial clustering of applications with noise) [Ester *et al.* (1996)] algorithm is a popular and well-cited density based clustering method proposed by Martin Ester, Hans-Peter Kriegel, Jorg Sander and Xiaowei Xu in 1996. In contrast to many other clustering methods that assume that clusters have convex shape, according to DBSCAN, clusters can form an arbitrary shape depending on the distance function that is used.

DBSCAN assumes that the instance space is built from high-density areas separated by low-density areas. Points that are within certain distance thresholds and comply with a density criterion are grouped together. An optional density criterion may require that the number of points within a certain radius should exceed a user-defined minimum.

There are two main parameters in DBSCAN algorithm: $MinPts$ and ϵ. These parameters are used to formally define dense areas. Based on these parameters' values, any point in the dataset is classified to one of the following labels: core points, density-reachable points and outliers.

A point p is said to be a core point if at least $MinPts$ points are within its $Eps - neighborhood$. The $Eps - neighborhood$ of a point p, denoted by $N_{Eps}(p)$, is defined as $N_{Eps}(p) = q \in X \mid dist(p, q) \leq \epsilon$, where X is the available dataset of points and $dist(p, q)$ indicates the distance function for two points p and q. All points in $N_{Eps}(p)$ are said to be directly density-reachable from p. The definition of directly density-reachable is symmetric for pairs of core points. However, it is not symmetric if one point is a border point and one point is a core point.

A point q is density-reachable from p if there is a path of points $p_1, \ldots, p_n$ with $p_1 = p$ and $p_n = q$, such that p_{i+1} is directly reachable from p_i. A point that is not reachable from any other point is considered an outlier by the algorithm. Figure 7.1 illustrates the various types of points. The point p is a core point. The point q is a border point which is density-reachable from p. The point o is an outlier.

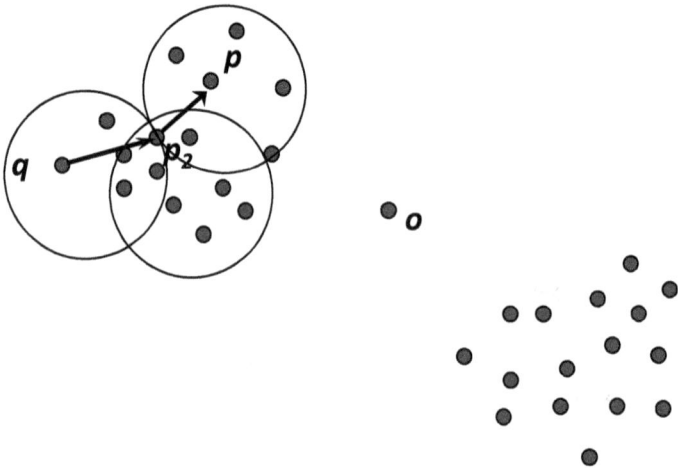

Fig. 7.1: Illustration of various types of points.

A cluster is a set of core points, that can be built by taking a core point, finding all of its neighbors that are core points, finding all of their neighbors that are core points, and so on. A cluster also consists of a set of non-core points, which are neighbors of a core point in the cluster but are not themselves core points. These points are referred to as border points.

Formally, a cluster C is a non-empty subset of X satisfying the following conditions:

Maximality: $\forall p, q \in X$: if $p \in C$ and q is density-reachable from p then $q \in C$.

Connectivity: $\forall p, q \in C$: p is density-connected to q.

Any cluster has at least $MinPts$ points in it. Further, any core point must be part of a cluster, by definition.

Generally speaking, the DBSCAN algorithm begins with arbitrarily selecting a point p. It then retrieves all points density-reachable from p with respect to ϵ and $MinPts$. If p is a core point, a cluster is created. The process continues until all of the points have been processed.

Figure 7.2 specifies the pseudo-code of DBSCAN algorithm. If a point is found to be a dense part of a cluster, then all points that are located within its neighborhood are added to the cluster by calling *expandCluster* function. A call of regionQuery(p, ϵ) returns the $Eps - neighborhood$ of p.

The computational complexity of DBSCAN is considered to be low, because it requires a linear number of range queries on the dataset and its results are fairly robust and therefore there is no need to run it several times.

7.2.1 *Running DBSCAN algorithm in R*

In this section, we illustrate how to use the DBSCAN algorithm that comes with the *dbscan* package [Hahsler (2015)] of R to cluster a sample of the Iris dataset. This implementation of DBSCAN uses a kd-tree data structure to reduce the computational cost. This makes it faster than the DBSCAN implementation that is included in *fpc* package of R.

Code 7.1 illustrates the usage of the dbscan algorithm. In line 1, we load the *dbscan* package. In lines 2-3, we prepare the iris dataset by focusing on the first four numeric dimensions. In order to perform the actual clustering, we need to set its two main parameters: *eps* and *minPts*. The *minPts* is often set close to the number of dimensions in the dataset. In order to determine the value of *eps* one can look at the knee in the distance plot. The plot is created in lines 5-7 using $k = 4$ (which the value of *minPts*) and the knee value is estimated as 0.5, as illustrated in Figure 7.3. On line 8, we perform the clustering, and in line 9, we print the textual summary of the model as presented in list 7.2. In this case, three clusters were detected by the algorithm, and 13 points are considered to be noise points. Figure 7.4 illustrates the resulting clustering in a scatter plot matrix. Each cluster is represented by a different symbol.

Require: X (instance set), ϵ, $MinPts$
Ensure: clusters
1: $C = 0$
2: **for all** $p \in X$ **do**
3:    **if** p has already been visited **then**
4:       continue to next point
5:    **else**
6:       mark p as visited
7:       $NeighborPts = regionQuery(\text{p}, eps)$
8:       **if** $\parallel NeighborPts \parallel \geq MinPts$ **then**
9:          $C = nextcluster$
10:          call expandCluster$(p, NeighborPts, C, \epsilon, MinPts)$
11:       **else**
12:          label P as NOISE
13:       **end if**
14:    **end if**
15: **end for**
16:
17: **function** expandCluster $(p, NeighborPts, C, \epsilon, MinPts)$
18: add p to cluster C
19: **for all** $q \in NeighborPts$ **do**
20:    **if** q has not been visited yet **then**
21:       mark q as visited
22:       $NeighborPtsQ = \text{regionQuery}(q, \epsilon)$
23:       **if** $\parallel NeighborPtsQ \parallel geMinPts$ **then**
24:          $NeighborPts = NeighborPts \cup NeighborPtsQ$
25:       **end if**
26:    **end if**
27:    **if** q is not yet member of any cluster **then**
28:       add q to cluster C
29:    **end if**
30: **end for**
31: **end function**
32:
33: **function** regionQuery (p, ϵ)
34:
35: **return** all points within P's eps-neighborhood (including P)
36: **end function**

Fig. 7.2: DBSCAN algorithm.

Listing 7.1: Using dbscan algorithm to cluster Iris dataset

```
1  library(dbscan)
2  data(iris)
3  iris <- as.matrix(iris[,1:4])
4
5  kNNdist(iris, k=4, search="kd")
6  kNNdistplot(iris, k=4)
7  ## the knee is around a distance of .5
8  res <- dbscan(iris, eps = .5, minPts = 4)
9  res
10 pairs(iris, pch = res$cluster + 1L)
```

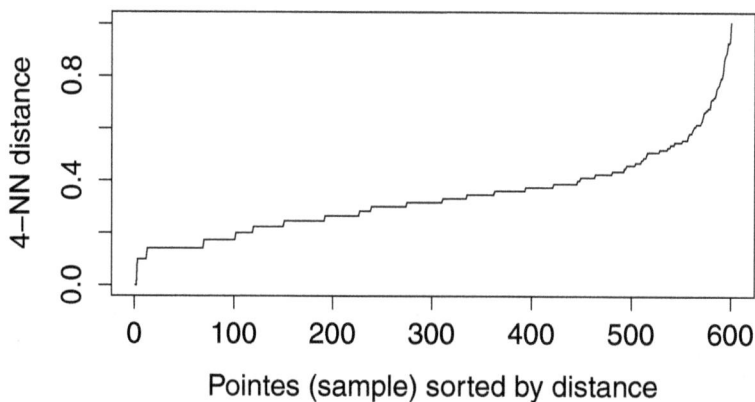

Fig. 7.3: knn distance plot.

Listing 7.2: Textual output of dbscan algorithm

```
1  DBSCAN clustering for 150 objects.
2  Parameters: eps = 0.5, minPts = 4
3  The clustering contains 3 cluster(s) and 13 noise points.
4
5   0  1  2  3
6  13 49 84  4
7
8  Available fields: cluster, eps, minPts
```

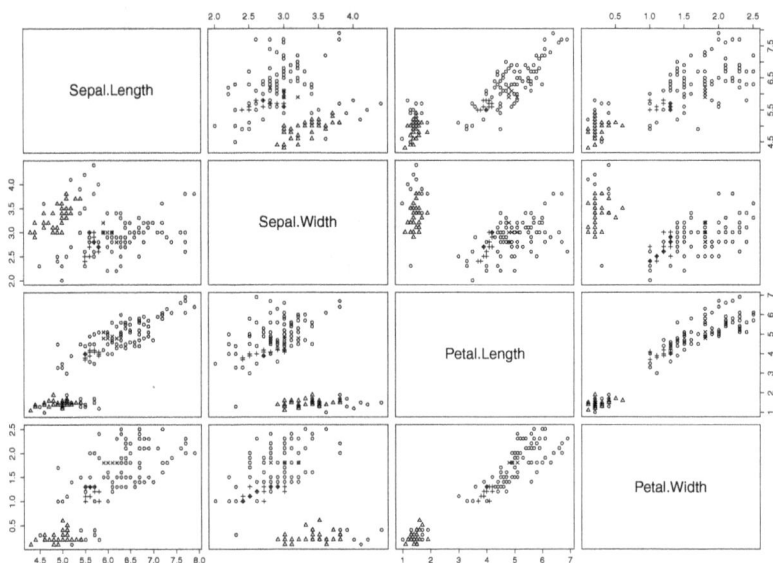

Fig. 7.4: Scatter plot matrix for dbscan clustering of Iris.

The function *predict* can be used to assign new data points to clusters. A point is assigned to a cluster if it is within the *eps* neighborhood of a member of that cluster. Noise points which cannot be assigned to a cluster are reported as cluster 0.

7.2.2 *Variations of DBSCAN and the OPTICS algorithm*

According to the original DBSCAN algorithm, border points are arbitrarily assigned to clusters. A revised version of DBSCAN, which is usually referred to as DBSCAN* [Campello *et al.* (2013)], labels all border points as noise points. This can be obtained by setting the parameter *borderPoints* in in *dbscan* to *FALSE*.

The key drawback of DBSCAN is that it requires substantial density reduction for detecting cluster borders. Moreover, DBSCAN cannot detect intrinsic cluster structures. The last drawback is resolved by a variation of DBSCAN called EnDBSCAN [Roy and Bhattacharyya (2005)].

The OPTICS (ordering points to identify the clustering structure) algorithm [Ankerst *et al.* (1999)] is a variation of DBSCAN which is capable to detect clusters in varying density data. For this purpose, the points are sorted such that points which are spatially closest become neighbours in the ordering. The pseudo-code of OPTICS is similar to that of DBSCAN, but it uses a priority queue to track unassigned points. In addition, OPTICS also provides a hierarchical clustering result and suggest performance improvement over DBSCAN. Moreover, the user is not required to tune the value of the meta-parameter ϵ and can simply set it to the maximum possible value. A careful value selection for ϵ can still helps in reducing the time complexity.

The *dbscan* package in R provides an implementation of the OPTICS original algorithm. This implementation has several hyper-parameters for controlling the algorithm. The *eps* parameter represents the upper limit for the neighborhood size, and it is used to reduce computational cost. The *minPts* parameter is usually set to be the number of features in the dataset. Code 7.3 illustrates the usage of OPTICS algorithm. In line 1, we load the *dbscan* package. In lines 2-3, we prepare the iris dataset by focusing on the first four numeric dimensions. In line 4, we perform the clustering. In line 6, we display the points ordered by the algorithm. The algorithm can report the reachability distance for each point in the dataset. The *plot* function can be used to produce the reachability plot, which specifies the reachability distance for every point. This is done in line 8, and the results are illustrated in Figure 7.5. In this plot, the valleys indicate clusters, and the hills represent points between clusters.

In lines 10-12, we identify the clusters and plot them. Points that have the same color are assigned to the same cluster. Black lines represent noise points. The clusters are grouped based on the value of the *eps_cl* parameter that serves as a threshold. The output vector *res$cluster* in line 14 holds the cluster assignment of the points. A cluster zero indicates noise points. Finally, in lines 16-20, we use the steepness threshold of *xi=0.01* to identify clusters hierarchically using the *Xi* method that is presented in the original OPTICS paper [Ankerst *et al.* (1999)]. Each level in the hierarchical structure represents the results that DBSCAN can produce with certain parameter values.

Listing 7.3: Using OPTICS algorithm to cluster Iris dataset

```
 1 library(dbscan)
 2 data(iris)
 3 iris <- as.matrix(iris[,1:4])
 4 res <- optics(x, eps = 0.4, minPts = 5)
 5 ### get points order
 6 res$order
 7 ### plot produces a reachability plot
 8 plot(res)
 9
10 ### identify clusters by cutting the reachability plot
11 res <- optics_cut(res, eps_cl = .3)
12 plot(res)
13 ## get cluster assignment
14 res$cluster
15
16 ### identify clusters using the hierarchically Xi method
17 res <- opticsXi(res, xi = 0.01)
18 plot(res)
19 # get Xi cluster structure
20 res$clusters_xi
```

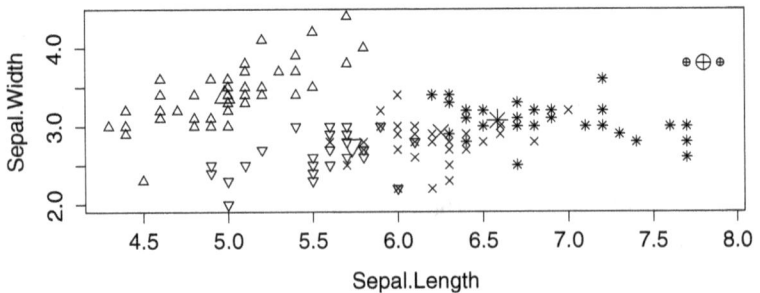

Fig. 7.5: Reachability plot of Iris.

7.3   Mean-shift

The main idea of mean-shift algorithm is to move each point to the dens-est area in its vicinity [Comaniciu and Meer (2002); Cheng (1995)]. The direction of the shift is determined using kernel density estimation. Over time, many points converge to a focal point of the density that can serve as clusters representatives, similarly to the centroids in k-means algorithm. Nevertheless in contrast to k-means mean-shift can detect arbitrary-shaped clusters. The computational complexity of mean-shift is considered to be higher than that of DBSCAN because it requires an expensive iterative procedure and density estimation.

Kernel density estimation aims to estimate the probability density at every point in space. Given a dataset x_i of points in d-dimensional space, and a kernel function f_K having bandwidth parameter h the density estimation is provided by:

$$f_K\left(\mathbf{x}\right) = \frac{1}{nh^d} \int_{i=1}^{n} K\left(\frac{\mathbf{x} - \mathbf{x_i}}{h}\right)$$ (7.1)

The kernel function should comply with the following two conditions:

$$\int K\left(\mathbf{x}\right) d\mathbf{x} = 1$$ (7.2)

$$K\left(\mathbf{x}\right) = K\left(|\mathbf{x}|\right) \qquad \forall \mathbf{x}$$ (7.3)

The first condition is needed to ensure a valid probability function. The second condition is needed to assume the symmetry of the input space. The most popular kernel function that satisfies the above condition is the Gaussian function:

$$K\left(\mathbf{x}\right) = \frac{1}{2\pi^{d/2}} e^{-\frac{1}{2}|\mathbf{x}|^2}$$ (7.4)

Given the kernel density estimation, we can now move points in the direction of locally increasing density. For this purpose, we can follow the direction of the function gradient. In particular, when the Kernel function is symmetric, the gradient takes the form:

$$\triangle f_k\left(\mathbf{x}\right) = \frac{2}{nh^{d+2}} \left(\sum_{i=1}^{n} -K'\left(\left|\frac{\mathbf{x}-\mathbf{x_i}}{h}\right|\right)\right) m\left(\mathbf{x}\right)$$ (7.5)

where $m\left(\mathbf{x}\right)$ is called the mean shift vector that moves the point toward the direction of increasing density, and K' represents the derivative of the kernel function:

$$m\left(\mathbf{x}\right) = \frac{\sum_{i=1}^{n} \mathbf{x}_i K'\left(\left|\frac{\mathbf{x}-\mathbf{x_i}}{h}\right|\right)}{\sum_{i=1}^{n} K'\left(\left|\frac{\mathbf{x}-\mathbf{x_i}}{h}\right|\right)} - \mathbf{x} \tag{7.6}$$

To conclude this section, we present the high-level pseudo-code in Figure 7.6.

Require: X (instance set), h, f_k
Ensure: clusters
 1: **repeat**
 2:    **for all** $p \in X$ **do**
 3:        Compute mean shift vector $m(p)$ for p using Eq. 7.6
 4:    **end for**
 5:    **for all** $p \in X$ **do**
 6:        Move each point: $p < --p + m(p)$
 7:    **end for**
 8: **until** convergence — i.e., points did not move too much in the last iteration.

Fig. 7.6: Mean shift algorithm.

The *msClustering* function in the *MeanShift* package performs clustering using the mean shift algorithm. It evaluates the modes of a kernel density estimator and assigns each point to one of the modes. All points that are assigned to the same mode constitute a cluster.

Code 7.4 demonstrates how to cluster the data with the Mean shift algorithm. Note that *msClustering* function assumes that the columns represent the points. Thus, we need to transpose the Iris data before performing the clustering (see Line 3). The h parameter should be set to a strictly positive value that represents the bandwidth in Equation (7.6). Lowering the value of h will produce a fine clustering (i.e., many small clusters), while increasing the value of h will result in a coarser clustering (i.e., few and large clusters). The *kernel* parameter is a string parameter that specifies the kernel function name and it can be one of the following: Gaussian (*kernel= "gaussianKernel"*), exponential (*kernel= "exponentialKernel"*), cubic (*kernel= "cubicKernel"*) or Epanechnikov (*kernel= "epanechnikovKernel"*).

Listing 7.4: Using Mean Shift algorithm to cluster Iris dataset

```
1  library(MeanShift)
2  data(iris)
3  iris <- t(as.matrix(iris[,1:4]))
4  res <- msClustering( iris, h=0.8 )
5  print(res)
6
7  plot( iris[1,], iris[2,], pch=res$labels*2, cex=1,
8          xlab="Sepal.Length", ylab="Sepal.Width" )
9  points( res$components[1,], res$components[2,],
10         pch=( 1:ncol( res$components ) )*2, cex=2)
```

The *labels* vector in the output consists of the points assignment. The *components* vector consists of the modes of the modes of the kernel density estimators. In lines 7-11 we plot the results using the first two dimensions (Sepal.Length and Sepal.Width) as illustrated in Figure 7.7. All points that are assigned to the same cluster are represented by unique symbol and the corresponding modes are represented by a larger symbol.

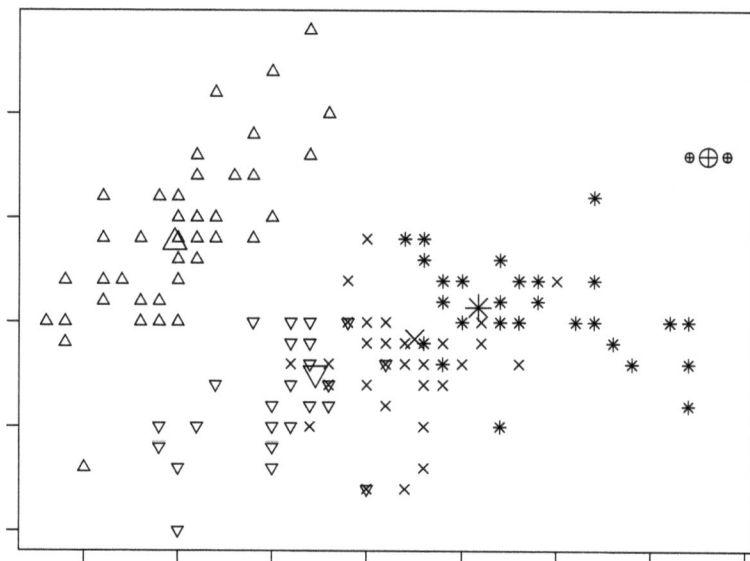

Fig. 7.7: Clustering Iris using mean-shift algorithm.

7.4   EM Clustering

The Expectation Maximization (EM) algorithm for clustering assumes that the data come from a mixture of probability distributions, each of which represents a different cluster. The EM algorithm follows an iterative approach which tries to find the parameters of the probability distribution by maximizing the likelihood. Each iteration consists of two main steps: E-Step (Expectation step) and M-Step (Maximization step). We begin by performing the E-step. The aim of this step is to estimate the probability of each point to belong to each cluster. The E-step is followed by the M-step, which aims to re-estimate the parameter vector of the probability distribution of each cluster. The algorithm terminates when the distribution parameters converge. The output of the algorithm is the distribution parameters and a soft assignment of points to clusters.

Each cluster j is represented by a Gaussian distribution. The Gaussian probability distribution is characterized by the mean (μ_j) and by the corresponding covariance matrix (Σ_j) We use the notation of $\theta(t)$ to indicate a parameter vector in time t such that:

$$\theta(t) = (\mu_j(t),\ \Sigma_j(t)),\ j = 1, \ldots, C \qquad (7.7)$$

During the initialization ($t = 0$), we can randomly choose initial values for the mean and the covariance matrix. Another option is to initialize θ with the clusters obtained by some clustering technique. We now go into the details of each of the two steps.

7.4.1   *E-step*

This step aims to(probabilistically) assign points to clusters under the current estimate of the model's parameters. In particular, we need to estimate the probability that a certain point x_i belong to a cluster j:

$$P^{(t)}\left(C_j \mid \mathbf{x_i}\right) = \frac{P^{(t-1)}\left(C_j\right) P^{(t-1)}\left(\mathbf{x_i} \mid C_j\right)}{\sum_{k=1}^{C} P^{(t-1)}\left(C_k\right) P^{(t-1)}\left(\mathbf{x_i} \mid C_k\right)} \qquad (7.8)$$

7.4.2   *M-step*

This step is responsible for estimating the parameters of the probability distribution of each class for the next step. We begin by calculating the probability of occurrence of each cluster:

$$P^{(t)}(C_j) = \frac{1}{N} \sum_{i=1}^{N} P^{(t)}(C_j \mid \mathbf{x_i}) \tag{7.9}$$

Then, the new mean of cluster j is calculated through the mean of all points weighted by their relevance:

$$\mu^{(t)}{}_j = \frac{\sum_{i=1}^{N} \mathbf{x_i} \; P^{(t)}(C_j \mid \mathbf{x_i})}{\sum_{i=1}^{N} P^{(t)}(C_j \mid \mathbf{x_i})} \tag{7.10}$$

Finally, we compute the new covariance matrix as follows:

$$\Sigma^{(t)}{}_j = \frac{\sum_{i=1}^{N} (\mathbf{x_i} - \mu^{(t)}{}_j)(\mathbf{x_i} - \mu^{(t)}{}_j)' \; P^{(t)}(C_j \mid \mathbf{x_i})}{\sum_{i=1}^{N} P^{(t)}(C_j \mid \mathbf{x_i})} \tag{7.11}$$

At the end of each iteration, we perform a convergence test which verifies if the difference between the current iteration and the previous iteration is lower than a predefined threshold defined by the user, namely if $\parallel \theta(t) - \theta(t-1) \parallel \le \epsilon$.

7.4.3 *Running EM algorithm in R*

The *mclust* package in R provides an implementation of the EM algorithm for parametrized Gaussian mixture models [Fraley and Raftery (2002, 2007); Fraley *et al.* (2012)]. There are several models supported in the package *mclust* as specified in Table 7.1.

Table 7.1: List of models supported in the package *mclust*.

Modelname	Description
"EII"	spherical, equal volume
"VII"	spherical, unequal volume
"EEI"	diagonal, equal volume and shape
"VEI"	diagonal, varying volume, equal shape
"EVI"	diagonal, equal volume, varying shape
"VVI"	diagonal, varying volume and shape
"EEE"	ellipsoidal, equal volume, shape, and orientation
"EVE"	ellipsoidal, equal volume and orientation
"VEE"	ellipsoidal, equal shape and orientation
"VVE"	ellipsoidal, equal orientation
"EEV"	ellipsoidal, equal volume and equal shape
"VEV"	ellipsoidal, equal shape
"EVV"	ellipsoidal, equal volume
"VVV"	ellipsoidal, varying volume, shape, and orientation

Code 7.5 illustrates how the EM algorithm can be used for clustering the Iris dataet. We begin by executing the m-step in line 3 for getting an estimation for the distribution's parameters. In line 6 we perform the EM algorithm using this estimation and setting the convergence threshold to *eps = 0.01*. Finally, at line 9 we project the raw points on dimensions 2 and 4 together with the parameters of the mixture model. The resulted graph is illustrated in Figure 7.8.

Listing 7.5: Using EM algorithm to cluster Iris dataset

```
1 library(mclust)
2 data(iris)
3 msEst <- mstep(modelName = "EEE", data = iris[,1:4],
4                z = unmap(iris[,5]))
5
6 res<-em(modelName = msEst$modelName, data = iris[,1:4],
7    parameters = msEst$parameters, control = emControl(eps =
       0.01))
8
9 coordProj( data=iris[,1:4], dimens=c(2,4), what = "errors",
      parameters=res$parameters, z=unmap(iris[,5]), truth =
      iris[,5])
```

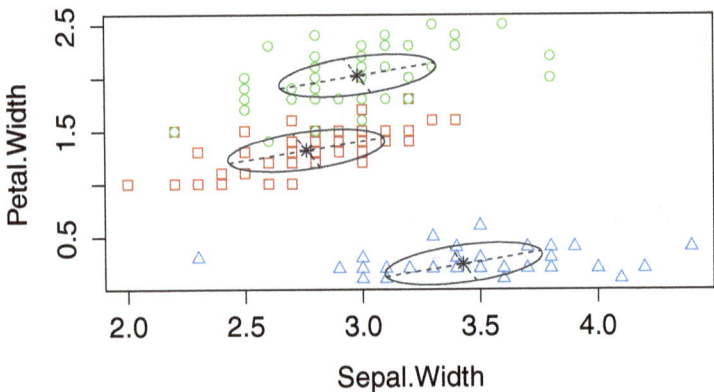

Fig. 7.8: EM clustering of iris.

Code 7.6 illustrates the usage of the *mclustBIC* function, which tries several models and number of clusters and then selects the top configuration based on the value of BIC (Bayesian information criterion). Figure 7.9 shows the obtained results. From the graph we can see that VEV model with 2 or 3 clusters provides the best BIC performance.

Listing 7.6: Using mclustBIC function to find the best clustering model for Iris dataset

```
1 library(mclust)
2 data(iris)
3 BIC = mclustBIC(iris[,1:4])
4 print(BIC)
5 plot(BIC)
```

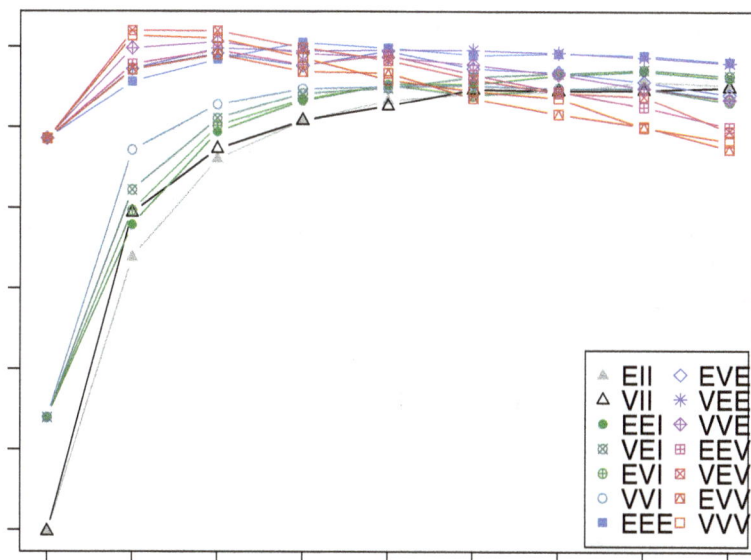

Fig. 7.9: BIC graph for various clustering of iris dataset.

In addition to the *mclust* package, the package *mixtools* also provides a set of functions for EM methods. In code we use *mvnormalmixEM* function for running the EM algorithm for mixtures of multivariate normal distributions. In this case we just provide the number of clusters ($k=3$) and the converge criterion (*epsilon*). On lines 4 and 5 we print the μ and Σ respectively. In line 6, we plot the clustering structure, and in line 7, we print the probability of assigning every point to each cluster.

Listing 7.7: Using mvnormalmixEM function to cluster Iris dataset

```
1 library(mixtools)
2 data(iris)
3 res <- mvnormalmixEM(iris[,c(1,3)], k=3,epsilon = 1e-02)
4 res$mu
5 res$sigma
6 plot(res, which=2)
7 res$posterior
```

7.5 Density Peak Clustering

Density-based clustering algorithms are a class of methods that identify clusters as dense regions of data points separated by sparser regions. Unlike partitioning methods such as k-means, density-based methods do not require the number of clusters to be specified a priori and are particularly effective in identifying clusters of arbitrary shape. One such method is the Density Peak Clustering (DPC) algorithm, which determines cluster centers based on the local density of points and their distance from points of higher density.

The Density Peak Clustering algorithm [Rodriguez and Laio (2014)] consists of several key steps: calculating the local density for each point, computing the distance from each point to its nearest point with higher density, and selecting cluster centers based on these two measures.

Below is the pseudo-code for the Density Peak Clustering algorithm. The algorithm takes as input a set of data points $\{x_i\}_{i=1}^n$ and a distance metric d. It outputs cluster labels for each data point. For each data point x_i, the local density ρ_i is calculated. This can be defined in various ways, such as counting the number of points within a certain distance or using a Gaussian kernel to weigh nearby points.

The distance δ_i for each point x_i is computed as the minimum distance to any other point x_j with a higher density ($\rho_j > \rho_i$). If x_i has the highest density, δ_i is the maximum distance to any other point. Points with both high local density ρ_i and large distance δ_i are selected as cluster centers. These points are typically far from other high-density points, making them natural choices for cluster centers. Each remaining point is assigned to the same cluster as its nearest neighbor with a higher density that has already been assigned to a cluster center.

The Density Peak Clustering algorithm can be implemented in R using the 'densityClust' package as illustrated in List 7.8. The densityClust

Require: Data points $\{x_i\}_{i=1}^{n}$, distance metric d
Ensure: Cluster labels for each data point
1: Compute the local density ρ_i for each point x_i
2: Compute the distance δ_i to the nearest point with higher density for each x_i
3: Select cluster centers based on high ρ_i and δ_i values
4: Assign each point x_i to the cluster of its nearest point with higher density that is a cluster center

Fig. 7.10: Density Peak Clustering.

Listing 7.8: Using Density Peak Clustering to cluster Iris dataset

```
1  library(densityClust)
2
3  irisDist <- dist(iris[,1:4])
4
5  irisClust <- densityClust(irisDist, gaussian=TRUE)
6  plot(irisClust) # Inspect clustering attributes to define
       thresholds
7
8  irisClust <- findClusters(irisClust, rho=2, delta=2)
9
10 plotMDS(irisClust)
```

function takes a distance matrix and, optionally a distance cutoff and calculates the values necessary for Density Peak Clustering. The actual assignment to clusters is done in a later step, based on user-defined threshold values. For this purpose, the findClusters function uses the supplied rho and delta thresholds to detect cluster peaks and assign the rest of the observations to one of these clusters.

Density Peak Clustering is a powerful method for identifying clusters of arbitrary shapes and densities. It excels in scenarios where clusters are not well-separated or have non-spherical shapes. However, it requires careful selection of parameters such as the distance threshold for density calculation and the cutoff for selecting cluster centers. Future research may focus on automating these parameter choices and extending the algorithm to handle high-dimensional and large-scale datasets more efficiently.

In summary, the Density Peak Clustering algorithm offers a robust approach to identifying clusters based on local density and distance, making it a valuable tool in the clustering toolbox.

7.6   Latent Class Analysis

Latent Class Analysis (LCA) is a model-based clustering technique that identifies unobserved (latent) subgroups within a population using categorical data. This probabilistic approach assumes that the population is a mixture of distinct classes, each with its own probability distribution. Unlike traditional clustering methods, LCA provides a statistical framework to determine the number of clusters and assess the uncertainty of class membership, offering a more robust and interpretable clustering solution.

LCA begins by defining a set of observed categorical indicators that are used to infer the latent classes. The core of LCA involves estimating the probability of class membership and the conditional probabilities of the observed indicators given the class membership. The goal is to maximize the likelihood of the observed data by iteratively adjusting the model parameters.

Input: Observed data matrix X with N samples and M categorical variables
Output: Estimated class probabilities $P(C_k|X)$ and conditional probabilities $P(X|C_k)$
Initialize: Randomly assign initial probabilities for class memberships $P(C_k)$ and conditional probabilities $P(X|C_k)$
repeat
    for each sample i in 1 to N **do**
        for each class k in 1 to K **do**
            Calculate posterior probability $P(C_k|X_i)$ using Bayes' theorem
        end for
    end for
    Update class membership probabilities $P(C_k)$ using the average posterior probabilities
    Update conditional probabilities $P(X|C_k)$ using the expectation maximization (EM) algorithm
until convergence criteria are met
Return: $P(C_k|X)$, $P(X|C_k)$

Fig. 7.11: Latent Class Analysis

The algorithm starts by initializing the class and conditional probabilities. For each sample, the posterior probability of belonging to each class is computed using Bayes' theorem. These probabilities are then averaged to update the class membership probabilities. The conditional probabilities are updated using the expectation-maximization (EM) algorithm, which iteratively adjusts the parameters to maximize the likelihood of the observed data. This process repeats until the algorithm converges, typically when changes in the probabilities fall below a predefined threshold.

To implement LCA in R, we can use the poLCA package, which provides functions to perform latent class analysis conveniently. First, install and load the poLCA package. Then, load your data into a data frame. Define the formula for the LCA model. The poLCA function runs the LCA model with the specified number of classes (in this case, 3). Finally, print and summarize the results to interpret the class probabilities and conditional distributions.

Listing 7.9: Using poLCA package to cluster election dataset

```
 1  # Install and load the poLCA package
 2  install.packages("poLCA")
 3  library(poLCA)
 4
 5
 6  data(election)
 7
 8  # Define the formula for the LCA model
 9  f2a <- cbind(MORALG,CARESG,KNOWG,LEADG,DISHONG,INTELG,
10             MORALB,CARESB,KNOWB,LEADB,DISHONB,INTELB)~PARTY
11
12  # Run the LCA with 3 classes
13  nes2a <- poLCA(f2a,election,nclass=3,nrep=5)
14
15  # Output the results
16  print(nes2a)
17  summary(nes2a)
18
19  #Ploting affinity
20  pidmat <- cbind(1,c(1:7))
21  exb <- exp(pidmat %*% nes2a$coeff)
22  matplot(c(1:7),(cbind(1,exb)/(1+rowSums(exb))),ylim=c(0,1),
        type="l",
23         main="Party ID as a predictor of candidate affinity
              class",
24         xlab="Party ID: strong Democratic (1) to strong
              Republican (7)",
25         ylab="Probability of latent class membership",lwd=2,
              col=1)
26  text(5.9,0.35,"Other")
27  text(5.4,0.7,"Bush affinity")
28  text(1.8,0.6,"Gore affinity")
```

Latent Class Analysis offers several advantages, such as providing a statistical framework for determining the number of clusters and handling mixed data types. However, it is computationally intensive and may struggle with very large datasets or high-dimensional data. Future research could focus on improving the scalability of LCA and integrating it with other clustering methods to enhance performance and applicability in various domains.

In conclusion, LCA is a powerful tool for identifying latent subgroups within heterogeneous populations, offering insights that can drive targeted interventions and personalized strategies.

7.7   Further Reading

- DBSCAN: [Andrade *et al.* (2013)], [Birant and Kut (2007)], [Borah and Bhattacharyya (2004)], [Chen *et al.* (2010)], [Dai and Lin (2012)], [Darong and Peng (2012)], [He *et al.* (2011)], [Jiang *et al.* (2011)], [Kisilevich *et al.* (2010)], [Kryszkiewicz and Lasek (2010)], [Patwary *et al.* (2012)], [Rong *et al.* (2004)], [Ruiz *et al.* (2007)], [Smiti and Elouedi (2012)], [Tran *et al.* (2013)], [Viswanath and Pinkesh (2006)], [Viswanath and Babu (2009)], [Xiaoyun *et al.* (2008)], [Yue *et al.* (2004)], [Zhou *et al.* (2000a)], [Zhou *et al.* (2000b)], [Zhou *et al.* (2000c)], [Zhou *et al.* (2012)]

- OPTICS: [Agrawal *et al.* (2016)], [Breunig *et al.* (1999)], [Breunig *et al.* (2000)], [Deng *et al.* (2015)], [Fu *et al.* (2015)], [Goyal *et al.* (2014)], [Kalita *et al.* (2007)], [Kriegel *et al.* (2003)], [Špitalský and Grendár (2013)], [Yu *et al.* (2016)]

- Mean-Shift: [Anand *et al.* (2014)], [Ardizzone *et al.* (2008)], [Bo *et al.* (2009)], [Carreira-Perpiñán (2006)], [Chu and Lee (2009)], [Georgescu *et al.* (2003)], [Luo and Khoshgoftaar (2004)], [Ozertem *et al.* (2008)], [Sharet and Shimshoni (2016)], [Subbarao and Meer (2006)], [Sutor *et al.* (2008)], [Toth *et al.* (2004)], [Tuzel *et al.* (2009)], [Wang *et al.* (2006)], [Wang *et al.* (2015)], [Wu and Yang (2007)], [Xiao and Liu (2010)], [Yuan *et al.* (2009)], [Yuan *et al.* (2012)], [Zhang *et al.* (2012)], [Zhou *et al.* (2008)]

- EM: [Dempster *et al.* (1977)], [McLachlan and Krishnan (2007)], [Borgelt and Kruse (2004)], [Bradley *et al.* (1998)], [El Assaad *et al.* (2016)], [Fayyad *et al.* (2001)], [Fazayeli *et al.* (2008)], [Guo *et al.* (2010)], [Gupta *et al.* (2010)], [Hidot and Saint-Jean (2010)], [Jin and Han (2011)], [Jollois and Nadif (2007)], [Jung *et al.* (2014)], [Kishor and Venkateswarlu (2016)], [Li and Bian (2009)], [Mustapha *et al.* (2009)], [Nasser *et al.* (2006)], [Safarinejadian *et al.* (2009)].

Chapter 8

Graph Clustering

8.1 Introduction

A graph is a common structure that consists of a set of vertices and a set of edges that are connected by edges. For example, an online social network such as Facebook is made up of a set of users (vertices) that are connected by friendship relations (edges). The goal of graph clustering is to partition the vertices of a graph into clusters based on vertices connectivity such that each cluster of vertices is densely connected internally and there are sparser connections between vertices that do not share the same cluster. Graph clustering has many applications. In social network analysis, graph clustering may be used to identify community structures. In bioinformatics, graph clustering is used to analyze protein interactions.

8.2 Graph Terminology

The graph is a structure made up of vertices (also known as nodes or points) and edges (also known as arcs or lines) that are connections between pairs of vertices. Graph clustering aims to group the vertices of a given input graph into clusters such that many vertices assigned to a particular cluster are connected by edges. Thus, a good clustering is made up of densely connected subgraphs that comprise as many as possible within-cluster edges while only a few of the edges connect vertices assigned to different clusters. A cluster in a graph is sometimes called a community.

Formally, a graph is an ordered pair $G = (V, E)$ with a set V of vertices together and a set E of edges. Every edge in E consists of a pair of vertices.

The number of edges in a given vertex v is the degree of v, denoted by $deg(v)$. A path is a sequence of distinct vertices $v_1, v_2, \ldots, v_n$ such that v_i is connected to v_{i+1} for all $i \in 1, 2, \ldots, n-1$. A cycle is the union of a path and an edge that connects v_n to v_i.

A graph is said to be a directed graph if the edges have a direction associated with them. A graph is said to be an undirected graph if edges have no orientation, i.e. edge (a, b) is identical to the edge (b, a). Many of the algorithms presented in this chapter assume that the underlying graph is an undirected graph. A graph is said to be a complete graph if every pair of distinct vertices is connected by an edge. A clique is a cluster of vertices such that its induced subgraph is complete i.e. every two distinct vertices in the clique are connected.

8.3 Affinity Propagation

Similarly to k-medoids algorithm, the affinity propagation algorithm [Frey and Dueck (2007)] utilizes the idea of exemplars. Examplar is an actual data point that is selected to represent a cluster of points. Recall that the k-medoids algorithm considers only a small subset of the points as potential exemplars, thus, it can provide a good clustering only if the initial subset of data points is close enough to the desired clustering. Contrary to k-medpods, the affinity propagation algorithm simultaneously considers all data points as candidate exemplars and therefore is not affected by the random selection of the initial subset. The algorithm gets the similarity matrix among all points as input. Real-value messages are exchanged among the points. The exchange value reflects the current affinity that one point has for choosing the second point as its exemplar. The values are updated iteratively until a high-quality set of exemplars is found.

There are two types of messages exchanged: responsibility and availability. Responsibility Message $r(\mathbf{x_i}, \mathbf{x_j})$ from point $\mathbf{x_i}$ to point $\mathbf{x_j}$ indicates the accumulated evidence for the adequacy of point $\mathbf{x_j}$ to serve as the exemplarfor for point $\mathbf{x_i}$ relative to other candidate exemplars for point $\mathbf{x_i}$. Availability message $a(\mathbf{x_i}, \mathbf{x_j})$ reflects the accumulated evidence for the appropriateness of point $\mathbf{x_i}$ to choose data point $\mathbf{x_j}$ as its exemplar. Both message values are initialized to zero. Then, in each iteration the responsibility values are updated according to:

$$r(\mathbf{x_i}, \mathbf{x_k}) \leftarrow s(\mathbf{x_i}, \mathbf{x_k}) - \max_{k' \neq k} \{a(\mathbf{x_i}, \mathbf{x_{k'}}) + s(\mathbf{x_i}, \mathbf{x_{k'}})\} \qquad (8.1)$$

where $s(\mathbf{x_i}, \mathbf{x_k})$ is a similarity measure, such as the similarity measures that were presented in Section 2.8. In particular, we can use the following measure as a default function:

$$s(\mathbf{x_i}, \mathbf{x_k}) = 1 - d(\mathbf{x_i}, \mathbf{x_k}) \qquad (8.2)$$

where $d()$ is the Gower distance.

Following the update of the responsibility values, the availability values are updated as follows:

$$a(\mathbf{x_i}, \mathbf{x_k}) \leftarrow \min \left(0, r(\mathbf{x_k}, \mathbf{x_k}) + \sum_{i' \notin \{i,k\}} \max(0, r(\mathbf{x_{i'}}, \mathbf{x_k})) \right) \quad i \neq k$$

$$(8.3)$$

$$a(\mathbf{x_k}, \mathbf{x_k}) \leftarrow \sum_{i' \neq k} \max(0, r(\mathbf{x_{i'}}, \mathbf{x_k})) \qquad (8.4)$$

The pseudocode of the Affinity Propagation algorithm is provided below. The input is a similarity matrix S where $S(i, k)$ indicates how well point i suits to be represented by point k. The algorithm begins by initializing responsibility and availability matrices to zero. Then, the algorithm runs iteratively until convergence. For each data point i and candidate exemplar k, update the responsibility $R(i, k)$. This update reflects the relative suitability of point k as the exemplar for point i. For each candidate exemplar k, update the self-availability $A(k, k)$. Then, for each data point i (except k), update the availability $A(i, k)$. This update ensures that the availability reflects the accumulated evidence for k being an exemplar. Finally, the algorithm outputs the identified exemplars and the corresponding clusters.

8.3.1 *Running affinity propagation algorithm in R*

In this section, we illustrate how to use the Affinity Propagation algorithm that comes with the *apcluster* package [Bodenhofer *et al.* (2011)] of R to cluster a sample of the Iris dataset (Figure 8.1). Code 8.1 begins by loading

1: **Input:** Similarity matrix s

2: Initialize responsibility $r(i,k) = 0$ and availability $a(i,k) = 0$

3: **while** not converged **do**

4:     **for** each data point i **do**

5:         **for** each candidate exemplar k **do**

6:             $r(i,k) \leftarrow s(i,k) - \max_{k' \neq k}\{a(i,k') + s(i,k')\}$

7:         **end for**

8:     **end for**

9:     **for** each candidate exemplar k **do**

10:        $a(k,k) \leftarrow \sum_{i' \neq k} \max\{0, r(i',k)\}$

11:        **for** each data point $i \neq k$ **do**

12:            $a(i,k) \leftarrow \min\{0, r(k,k) + \sum_{i' \notin \{i,k\}} \max\{0, r(i',k)\}\}$

13:        **end for**

14:    **end for**

15: **end while**

16: **Output:** Exemplars and clusters

Fig. 8.1: Affinity propagation.

Listing 8.1: Using affinity propagation algorithm to cluster Iris dataset

```
1 library(apcluster)
2 set.seed(2)
3 irisSample <- iris[sample(nrow(iris),64),1:4]
4 sim <- negDistMat(irisSample,r=2)
5 clust <- apcluster(sim)
6 show(clust)
7 plot(clust,irisSample)
```

the *apcluster* package. In lines 2-3, we prepare the Iris dataset by focusing on the first four numeric dimensions. In line 4, we compute the similarity matrix using the standard negative squared distances. In line 5, we execute the algorithm and get the clustering results, which are then printed in line 6. The textual summary of the model is presented in list 8.2. Figure 8.2 illustrates the resulting clustering in a scatter plot matrix.

Listing 8.2: Textual output of affinity propagation algorithm

```
 1 APResult object
 2
 3 Number of samples      =   12
 4 Number of iterations   =   127
 5 Input preference       =   -3.555
 6 Sum of similarities    =   -3.96
 7 Sum of preferences     =   -10.665
 8 Net similarity         =   -14.625
 9 Number of clusters     =   3
10
11 Exemplars:
12     28 138 67
13 Clusters:
14     Cluster 1, exemplar 28:
15         28 25 19 34
16     Cluster 2, exemplar 138:
17         105 138 137 78 141
18     Cluster 3, exemplar 67:
19         85 120 67
```

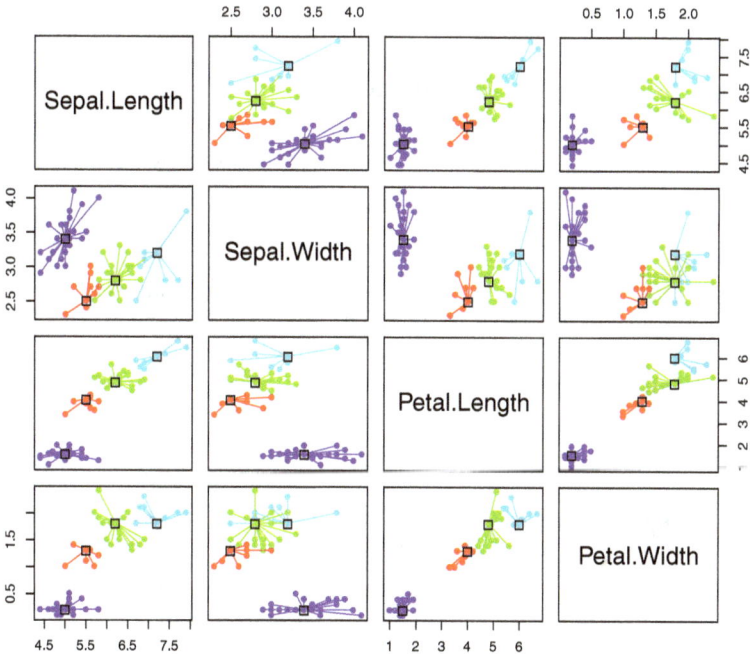

Fig. 8.2: Affinity propagation plot for Iris dataset.

8.3.2 *Conclusions, advantages, limitations, and future research*

Affinity Propagation (AP) is a clustering algorithm that identifies exemplars among data points and forms clusters by assigning each point to its closest exemplar. This algorithm is particularly notable for its ability to handle non-convex clusters and does not require the number of clusters to be specified beforehand, making it a robust and flexible choice for unsupervised learning.

AP operates by exchanging real-valued messages between data points until a good set of exemplars and corresponding clusters could be detected. The messages update in a way that reflects the evidence for one point being the exemplar of another.

Affinity Propagation offers several advantages:

- It does not require the number of clusters to be pre-determined.
- It handles non-convex clusters well.
- It identifies exemplars, providing interpretable clustering results.

However, it has some limitations:

- It can be computationally intensive for large datasets.
- The choice of similarity function can significantly affect results.
- Sensitivity to hyperparameters, particularly the preference parameter.

Future research directions include:

- Enhancing the efficiency of the algorithm for large datasets.
- Developing methods to automatically tune hyperparameters.
- Extending the algorithm to handle streaming data.

8.4 K-Cores

The K-cores clustering is an algorithm for simplifying graph topology by pruning all the vertices (with their respective edges) with a degree less than k [Batagelj and Zaveršnik (2011)]. The notion of k-cores was proposed in 1983 by Seidman [Seidman (1983)], in an attempt to decompose a large graph into smaller and more manageable sub-graphs. The algorithm is executed in a cascade mode such that the pruning of certain vertices may eventually cause the pruning of additional vertices. In particular, the new degree of a vertex v that has n neighbors with degree less than k, is $deg(v) - n$ where $deg(v)$ is the original degree of vertex v. Moreover, vertex v will be pruned if $deg(v) - n < k$.

Formerly, the notion of generalized k-core is defined as follows. Given a graph $G = (V, E)$ and a real-valued vertex property function $f(v, S, G)$ where $v \in V$ and $S \subseteq V$, then a subset $C \subseteq V$ is a generalized k-core for f if C is a maximal set such that $f(v, C, G) \geq k; \forall v \in C$. The function f represents by default the degree measure of the nodes. In such case, every set member in C is linked to at least k other members of C. Nevertheless, other functions can be used as well. For example, the function f can be refined to count the number of paths between any two pairs of vertices. This can be used to obtain a k-component sub-graph, a maximal graph that cannot be disconnected by fewer than k nodes [Myrvold (1992)].

8.4.1   *Running K-cores clustering in R*

In this section, we illustrate how to use the *kcores* function that comes with the *sna* package [Butts (2010)] of R to produce degree-based k-cores. The input data is an adjacency matrix representing the graph. The output is a vector containing the maximum core membership for each vertex based on the selected degree measure.

Code 8.3 begins by loading the *sna* package and *cluster*, which is needed for calculating the distances between the points in the Iris dataset. In line 4, we prepare the Iris dataset by focusing on the first four numeric dimensions. In line 5, we compute the dissimilarity matrix using Euclidean distance. In line 6, we convert the distance matrix into an adjacency matrix by constructing an edge between any two points whose distance is less than 1. In line 7, we execute the algorithm using the default parameters. The core numbers for each point are printed in line 8. The textual summary of the model is presented in list 8.4. In line 10, we plot the graph so that the color of each vertex is determined by its core value. Figure 8.3 illustrates this graph.

Listing 8.3: Using K-cores algorithm to cluster Iris dataset

```
1 library(sna)
2 library(cluster)
3 set.seed(2)
4 irisSample <- iris[sample(nrow(iris),25),1:4]
5 DisMat <- as.matrix(daisy(irisSample, metric = "euclidean",
      stand = FALSE))
6 AdjMat <- DisMat<1
7 kc<-kcores(AdjMat)
8 show(kc)
9 #Plot the result
10 gplot(AdjMat,vertex.col=kc)
```

Listing 8.4: Textual output of K-cores clustering algorithm

1	28	105	85	25	138	137	19	120	67	78	141	34	149	147	56
		116	131	31	59	10								
2	12	10	4	12	10	10	8	4	4	10	10	8	10	6	4
		10	2	12	6	12								
3	87	50	108	20	44									
4	6	12	2	12	12									

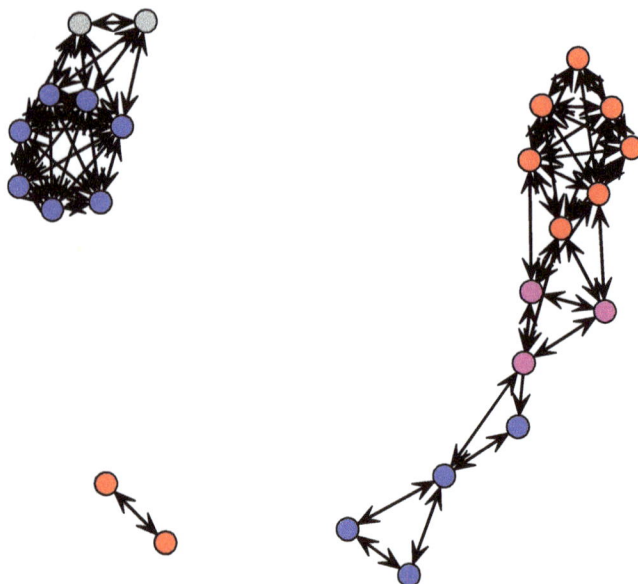

Fig. 8.3: K-cores clustering plot for Iris dataset.

8.5 The Igraph Package

The *igrpah* package is a well-known package for creating, manipulating and analyzing graphs. The package is written in C, but it provides API in both R and Python. The main advantage of *igrpah* is its ability to handle large networks efficiently. In this section, we will present and demonstrate the basic functionality that *igrpah* provides.

8.5.1 *Creating graphs*

There are many ways to create graphs using the *igrpah* package. We begin by generating a graph from scratch. Code 8.5 begins by loading the *igrpah* package. In line 2, we create an empty undirected graph with five vertices by setting the parameter *n* to *5* and *directed* to *false*. In line 3, we add a few edges. Each edge is provided as a pair of vertices to be connected. Specifically, we connect vertex *1* to vertices *3,4,5* and create a loop in the graph by connecting vertex *1* to itself. In line 4, we set the color attribute of the edges, and finally, in line 5, we plot the graph.

Listing 8.5: Creating a graph from scratch using *igrpah* package.

```
1 library(igraph)
2 myGraph<-graph.empty(n=5, directed=FALSE)
3 myGraph<-add_edges(myGraph, c(1,3, 1,4, 2,2, 1,5))
4 myGraph<-set_edge_attr(myGraph, "color", value = "blue")
5 plot(myGraph)
```

The graph package has many other functions for creating graphs. Here, we specify the most common:

- graph.full that creates a full graph in which each pair of vertices has an edge connecting them.
- graph.tree that creates a regular tree graph by obtaining the number of vertices and number of children.
- graph.star that creates a star graph, in which there is a center vertex that is connected to all other vertices.
- graph.lattice that is used to create lattices of arbitrary dimensions. We can either specify the number of vertices in each dimension of the lattices or set the number of dimensions of the lattice and a constant size of each dimension.

- graph.ring that creates a one-dimensional circular graph.
- graph.adjacency that creates a graph from an adjacency matrix.

Code 8.6 demonstrates the abovementioned functions for creating a graph. Note that to create a graph from the adjacency matrix (line 9), we first need to set the matrix (lines 7-8). Finally, in line 10, we define an array of plot placeholders that are actually being plotted in lines 11 to 16. The resulting plot array is presented in Figure 8.4.

Listing 8.6: Generating graphs using different methods.

```
1 library(igraph)
2 myTree<-graph.tree(8, children = 2)
3 myFull<-graph.full(5, directed = FALSE, loops = FALSE)
4 myStar<-graph.star(5, center = 1)
5 myLattice<-graph.lattice(dimvector = c(2,3,5))
6 myRing<-graph.ring(5, directed = FALSE, mutual = FALSE,
    circular=TRUE)
7 adjm <- matrix(sample(0:5, 25, replace=TRUE,
8                prob=c(0.5,0.05,0.05,0.1,0.1,0.2)), nc=5)
9 myAdj <- graph.adjacency(adjm, weighted=TRUE)
10 par(mfrow=c(3,2),mai=c(0,0,0,0),mex=c(0,0,0,0))
11 plot(myTree)
12 plot(myFull)
13 plot(myStar)
14 plot(myLattice)
15 plot(myRing)
16 plot(myAdj)
```

8.5.2 *Centrality measures*

The *igraph* package provides methods for calculating various centrality measures. These measures are very useful for detecting clusters in the graph because they characterize the vertices and the edges of the graph. In the following sub-sections, we review the most common measures.

8.5.2.1 *Degree*

The degree centrality of a given vertex v (denoted as $deg(v)$) simply counts the number of adjacent edges. For the directed graph, we can differentiate between two separate measures of degree centrality: outdegree and indegree. As the name implies, outdegree (denoted as $deg^+(v)$) counts the number of edges that the vertex directs to other vertices, while indegree (denoted as $deg^-(v)$) refers to the number of edges directed toward

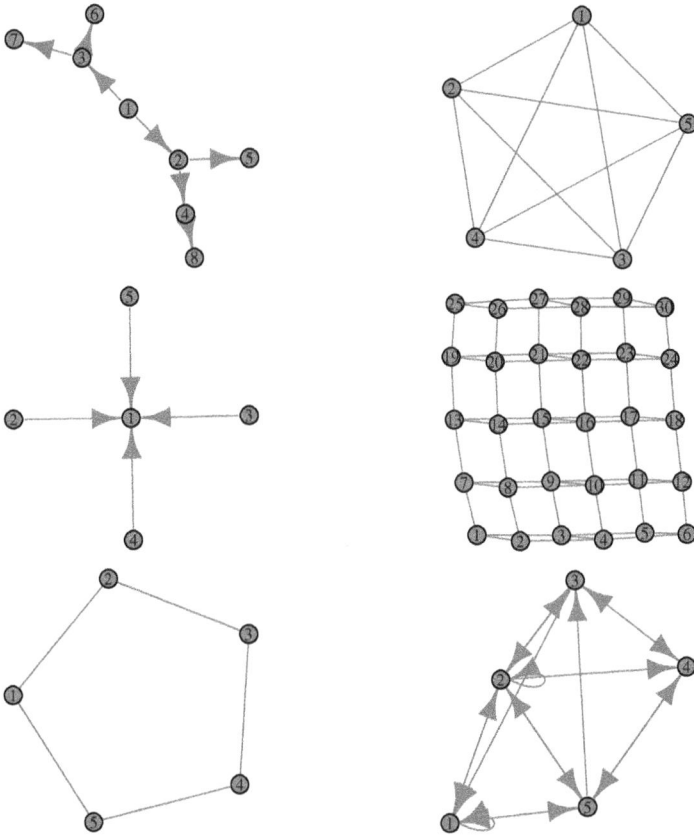

Fig. 8.4: The various graphs that were created by code 8.6.

the vertex. A vertex with zero indegree is called *source*, and a vertex with zero outdegree is called *sink*.

In a social network, a high in-degree value is often associated with popular entities (e.g., celebrities), while high outdegree values are often associated with gregariousness.

8.5.2.2 *Betweenness*

The betweenness score of a vertex measures the centrality of the vertex by counting the number of shortest paths traversing through this vertex. A shortest path between a pair of vertices is defined as the path with the minimal number of edges.

Formaly, the betweenness score of a vertex v is defined as:

$$g(v) = \sum_{i \neq v \neq j} \frac{\sigma_{ij}(v)}{\sigma_{ij}} \tag{8.5}$$

where σ_{ij} is the number of shortest paths that goes from vertex i to vertex j and $\sigma_{ij}(v)$ is the total number of those paths that pass through vertex v.

8.5.2.3   *Closeness*

The closeness centrality of a given vertex counts the total number of steps required to reach all other vertices from the given vertex. Specifically, the closeness centrality is defined by the inverse of the average length of the shortest paths from a given vertex v to all the other vertices in the graph:

$$C(v) = \frac{1}{\sum_i d(i, v)} \tag{8.6}$$

where $d(i, v)$ is the distance between vertices i and v. To allow comparisons of graphs of different sizes, we recommended to nomalize the closeness value by multipling it by $N - 1$, where N is the number of vertices in the graph.

The calculations of the above centrality measures are demonstrated in Code 8.7. We begin by randomly generating a directed graph with ten vertices and an edge probability of 0.3. In lines 4-6, we calculate various degree measures (indegree, outdegree and regular degree). Then, we calculate the closeness and betweenness (lines 7-8). Finally, in line 9, we plot the graph (see Figure 8.5). Next to each vertex, we display its betweenness score. As expected, sources and sinks are expected to have a zero betweenness centrality.

Listing 8.7: Demonstrating of centrality measures calculation.

```
1 library(igraph)
2 set.seed(73)
3 g <- sample_gnp(n=10, p=0.3, directed = TRUE)
4 d1=degree(g, mode="in")
5 d2=degree(g, mode="out")
6 d3=degree(g, mode="all")
7 c=closeness(g, mode="all")
8 b=betweenness(g)
9 plot(g,vertex.label.dist =1.5,vertex.label.cex=2,vertex.
      label=round(b,2))
```

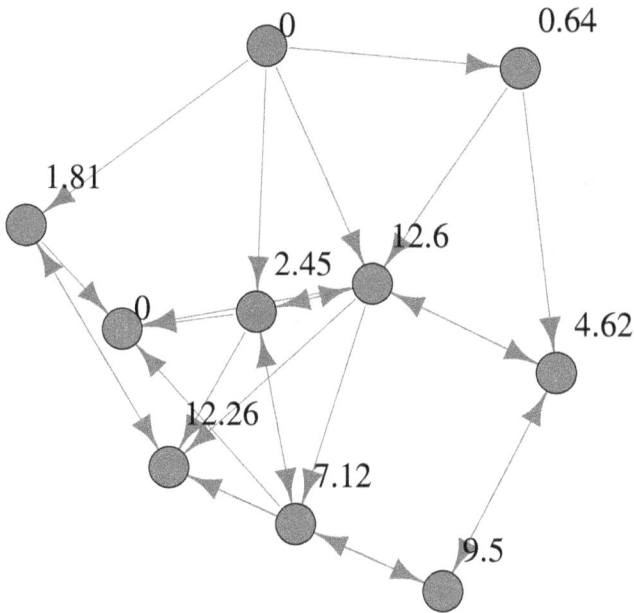

Fig. 8.5: Graph generated by code 8.7.

8.5.3 Community structure detection based on edge betweenness

An edge with a high betweenness score may indicate that this edge connects two communities because many shortest paths from vertices in one community to vertices in the other community must go through it. Thus, for detecting communities, we can greedily remove the edge with the highest betweenness score. After removing an edge, we recalculate the edge betweenness scores and remove the edge with the highest score and so forth until all vertices are disconnected. The result of this process is a dendrogram where the root represents the entire graph, and the leaves store the individual vertices. This algorithm can be performed using the *cluster_edge_betweenness* function.

Code 8.8 demonstrates the use of *cluster_edge_betweenness* function. We begin by creating an artificial graph using *barabasi.game* procedure. The procedure begins with a single vertex. Then, in every iteration, a new vertex is added, and it is connected to previous vertices (e.g., i) according to the following probability:

$$P_i \sim k_i^\alpha + a \tag{8.7}$$

where k_i is the current total number of edges of vertex i which were not created by i itself. The parameters α and a are set by the power and zero.appeal arguments correspondently.

In line 4 the *cluster_edge_betweenness* algorithm is executed, and its corresponding output is plotted in two ways: as a dendrogram in line 5 (see the result in Figure 8.6) and as a graph in line 6 (see Figure 8.7). Note that in both plots, the main communities are highlighted in different colors.

Listing 8.8: Demonstrating *cluster_edge_betweenness* algorithm.

```
1 library(igraph)
2 set.seed(73)
3 g <- barabasi.game(20,m=2,power=1,zero.appeal=1)
4 eb <- cluster_edge_betweenness(g)
5 plot_dendrogram(eb)
6 plot(eb,g)
```

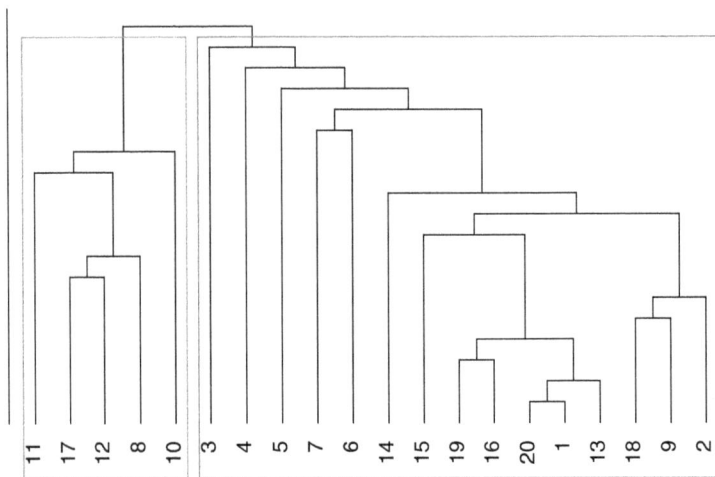

Fig. 8.6: A dendrogram created by code 8.8.

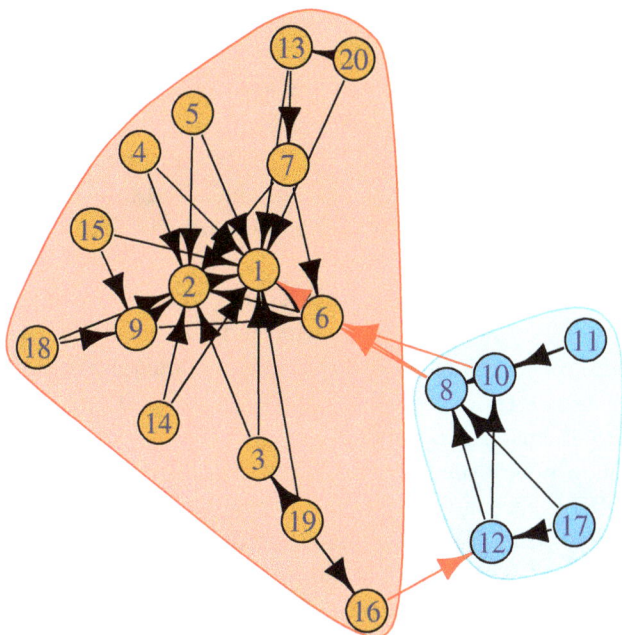

Fig. 8.7: A clustering created by code 8.8.

8.6 CHAMELEON

Many existing clustering algorithms make assumptions about the shape, size, or density of the clusters that limit their effectiveness on more complex data sets. The CHAMELEON algorithm, proposed by Karypis et al. [Karypis *et al.* (1999)], aims to address these limitations by dynamically modeling the clusters during the clustering process.

The key idea behind CHAMELEON is to measure the similarity between two clusters based on a dynamic model that considers both the inter-connectivity and closeness of the clusters relative to the internal characteristics of the clusters themselves. This allows CHAMELEON to discover arbitrary-shaped clusters of varying density and size.

CHAMELEON operates in two phases:

(1) Find initial sub-clusters by partitioning the similarity graph of the data using a graph partitioning algorithm.

(2) Merge these sub-clusters hierarchically based on a dynamic model of relative inter-connectivity and closeness.

The algorithm represents the data as a sparse k-nearest neighbor graph, where nodes are data points and edges connect each point to its k-nearest neighbors, weighted by similarity. Finding the initial sub-clusters is treated as a graph partitioning problem to minimize the edge-cut while balancing the sub-cluster sizes.

In the merging phase, CHAMELEON uses two novel measures of cluster similarity - relative inter-connectivity and relative closeness - to decide which clusters to merge at each step. This dynamic modeling of clusters based on their own characteristics allows CHAMELEON to be effective for a wide range of cluster shapes, densities, and sizes.

The pseudocode of the CHAMELEON algorithm is given in Figure 8.8. The CHAMELEON algorithm takes a data set X, the number of nearest neighbors k to use for the similarity graph, and the minimum desired sub-cluster size $MinSize$ as input.

It first constructs the k-nearest neighbor graph G from X, where nodes represent data points and there is an edge between two nodes if one data point is among the k-nearest neighbors of the other, weighted by their similarity (line 1).

In the first phase (line 2), it partitions this graph G into a set of sub-clusters $SubClusters$ such that the edge-cut between sub-clusters is minimized while approximately balancing the sub-cluster sizes to be at least $MinSize$. This is accomplished using a graph partitioning algorithm.

The second phase (lines 4-8) is an agglomerative hierarchical merging process that starts with the $SubClusters$ and iteratively merges the two most similar clusters until only one cluster remains.

The core operation is selecting the pair of clusters C_i and C_j to merge at each step (line 5) by maximizing the product of their relative inter-connectivity $RI(C_i, C_j)$ and relative closeness $RC(C_i, C_j)$. These two measures dynamically model the similarity between clusters based on internal cluster properties, as described below.

After merging C_i and C_j into a new cluster $NewCluster$ (line 6), the original clusters are removed from $Clusters$ and the new merged cluster is added (line 7).

This process repeats until only one cluster remains, which is returned as the final clustering (line 9).

Require: Data set X, number of nearest neighbors k, minimum sub-cluster size $MinSize$

1: Construct the k-nearest neighbor graph G from X
2: $SubClusters \leftarrow PartitionGraph(G, MinSize)$
 Phase 1: Find initial sub-clusters
3: $Clusters \leftarrow SubClusters$
4: **while** $|Clusters| > 1$ **do**
5: $\quad C_i, C_j \leftarrow \text{argmax}_{C_i, C_j \in Clusters} RI(C_i, C_j) \cdot RC(C_i, C_j)$
 Select pair of clusters to merge based on relative inter-connectivity (RI) and closeness (RC)
6: $\quad NewCluster \leftarrow Merge(C_i, C_j)$
7: $\quad Clusters \leftarrow Clusters \setminus \{C_i, C_j\} \cup \{NewCluster\}$
8: **end while**
9: **return** $Clusters$

Fig. 8.8: CHAMELEON clustering algorithm.

Here is how the relative inter-connectivity $RI(C_i, C_j)$ between clusters C_i and C_j is defined:

$$RI(C_i, C_j) = \frac{|EC_{C_i,C_j}|}{(|EC_{C_i}| + |EC_{C_j}|)/2}$$

where EC_{C_i,C_j} is the sum of edge weights between points in C_i and C_j (the "edge cut" separating the clusters) while EC_{C_i} and EC_{C_j} are the respective min-cut bisector values for C_i and C_j (the weighted edge-cut values for partitioning each cluster into two roughly equal parts).

So $RI(C_i, C_j)$ measures the absolute inter-connectivity between C_i and C_j relative to the internal inter-connectivity within each cluster. This normalizes for differences in cluster tightness and density.

The relative closeness $RC(C_i, C_j)$ between clusters C_i and C_j is defined as:

$$RC(C_i, C_j) = \frac{S_{EC_{C_i,C_j}}}{\frac{|C_i|}{|C_i|+|C_j|}S_{EC_{C_i}} + \frac{|C_j|}{|C_i|+|C_j|}S_{EC_{C_j}}} \tag{8.8}$$

where $S_{EC_{C_i,C_j}}$ is the average weight of edges between points in C_i and C_j, $S_{EC_{C_i}}$ and $S_{EC_{C_j}}$ are the average weights of edges in the respective min-cut bisectors for C_i and C_j, $|C_i|$ and $|C_j|$ are the sizes (number of points) in each cluster.

So $RC(C_i, C_j)$ measures how close the clusters are relative to their internal closeness, where closeness is based on the average inter-cluster similarities along the cluster boundaries rather than distance between centroids or medoids.

By considering both relative inter-connectivity and closeness, CHAMELEON can balance separating tight clusters and merging nearby clusters of varying density without being overly influenced by outliers or noise.

These two phases of finding initial sub-clusters and then dynamically merging them based on the RI and RC criteria allow CHAMELEON to effectively discover arbitrarily shaped clusters in the data.

The CHAMELEON algorithm takes a unique approach to clustering by dynamically modeling the similarity between clusters during the merging process, rather than using a static model or only considering certain cluster properties like density or inter-connectivity.

In the first phase, it efficiently finds an initial set of tight sub-clusters by treating it as a graph partitioning problem on the k-nearest neighbor similarity graph.

The real novelty lies in the second merging phase that repeatedly identifies the most similar pair of clusters C_i and C_j to merge together based on maximizing the product of their relative inter-connectivity $RI(C_i, C_j)$ and relative closeness $RC(C_i, C_j)$.

The $RI(C_i, C_j)$ measure compares the absolute inter-connectivity between C_i and C_j (summed edge weights along the cut between them) to the averaged internal inter-connectivity within each cluster (measured by their min-cut bisector values). This accounts for differences in cluster tightness so that relative differences matter more than absolute differences.

The $RC(C_i, C_j)$ measure captures the relative closeness of the clusters by comparing the averaged similarity between points near the boundaries of C_i and C_j to the averaged internal closeness within each cluster (based on the edge weights in their min-cut bisectors). This offers a noise-tolerant way to assess cluster proximity without being misled by outliers.

By combining these two measures into a single criterion $RI(C_i, C_j) \cdot RC(C_i, C_j)$, CHAMELEON selects cluster pairs that are both well interconnected and close together, relative to their own internal characteristics. This dynamic modeling allows it to be effective for clusters of arbitrary shape, size, and density.

Overall, the two-phase approach of first finding tight sub-clusters and then dynamically merging them based on the RI and RC criteria enables

CHAMELEON to discover the natural, genuine clusters present in complex data sets with varying properties.

CHAMELEON is a powerful clustering algorithm that offers several key advantages over traditional approaches:

(1) It can discover arbitrarily shaped clusters of varying size and density within the same dataset.
(2) The two-phase approach allows it to scale well to large data sets.
(3) It is robust to outliers and noise by focusing on relative similarities.
(4) It does not require apriori specification of parameters like the number of clusters.

Some potential limitations of CHAMELEON include:

(1) It requires computing the full similarity matrix/nearest neighbor graph, which can be expensive for very large datasets.
(2) Distance/similarity measures may not be meaningful for high dimensional data.
(3) Setting the parameters k, minSize requires some trial and error.

Overall, CHAMELEON is a highly effective and versatile clustering technique, especially well-suited for datasets exhibiting wide variation in cluster properties.

Future research directions include:

(1) Extending the dynamic modeling approach to other clustering paradigms like density-based or grid-based.
(2) Developing efficient approximations to the k-nearest neighbor graph construction.
(3) Semi-supervised variants that can incorporate user guidance or constraints.
(4) Automating the parameter selection process through techniques like model order estimation.

8.7 The CACTUS Algorithm for Clustering Categorical Data

The key idea behind the CACTUS (Categorical Clustering Using Summaries) algorithm is to summarize the dataset in a way that allows identifying clusters of categorical data efficiently. It introduces a novel definition of a cluster for categorical data by generalizing the concept of dense regions from numeric data. The CACTUS algorithm operates in three phases:

(1) **Summarization phase:** Compute summaries of the data that capture the co-occurrence of attribute values.
(2) **Clustering phase:** Use the summaries to efficiently identify candidate clusters.
(3) **Validation phase:** Verify which of the candidate clusters satisfy the minimum support requirement on the original data.

By first summarizing the data into a compact representation that fits in memory, CACTUS avoids the need to repeatedly scan the full dataset. This allows it to discover clusters very quickly compared to previous approaches.

Below is a detailed walkthrough of how the CACTUS algorithm works, including pseudocode and explanations. The overall algorithm consists of three main parts: summarization, clustering, and validation.

The first phase computes two types of summaries from the data: inter-attribute summaries that capture co-occurrences between pairs of attribute values from different attributes, and intra-attribute summaries that summarize similarities between values of the same attribute.

To compute the inter-attribute summaries Σ_{IJ}, we iterate through all pairs of attributes A_i, A_j. For each pair, we initialize $c(a_i, a_j)$ to zero for all pairs of attribute values (a_i, a_j) from the respective domains. We then scan through the dataset D once, incrementing the appropriate $c(a_i, a_j)$ counts for each tuple t.

After scanning D, we determine which pairs (a_i, a_j) are strongly connected, meaning their co-occurrence count $c(a_i, a_j)$ exceeds the expected count $\alpha \frac{|D|}{|D_i||D_j|}$ under an assumption of attribute independence. Here, $\alpha > 1$ is a parameter specifying how much larger the observed co-occurrence must be compared to the expected count. The set of all strongly connected pairs $(a_i, a_j, c(a_i, a_j))$ forms the inter-attribute summary σ_{ij} for A_i, A_j.

To compute the intra-attribute summaries Σ_{II}, we iterate through the same pairs of attributes A_i, A_j. For a given pair (A_i, A_j), we define the similarity between two values $a_i, a_i' \in D_i$ with respect to A_j as the number of values $a_j \in D_j$ for which both (a_i, a_j) and (a_i', a_j) are strongly connected pairs in σ_{ij}. The set of all such similarities is the intra-attribute summary σ_{ii}^j for attribute A_i with respect to A_j.

The key ideas are:

(1) The summaries only track strongly connected pairs, filtering out noise.
(2) They can be computed efficiently with two scans over the data.
(3) For categorical attributes with small domain sizes, the summaries fit in memory.

In the second phase, CACTUS uses the summaries to efficiently find clusters in the data by exploiting the following properties:

(1) A cluster over n attributes induces a 2-cluster over any pair of those attributes. The cluster projections over individual attributes can be found by intersecting these 2-clusters.
(2) Any subcluster of a cluster also qualifies as a cluster in its own right.

The algorithm first computes cluster projections for each attribute, and then synthesizes full n-dimensional candidate clusters from these projections in a level-wise manner.

For each attribute A_i, CACTUS computes the cluster projections of 2 clusters between A_i and every other attribute A_j.

The idea is to first find all small sets $S \subseteq D_i$ of size $\leq \delta$ whose values form cliques in the intra-attribute summary σ_{ii}^j. We call these the distinguishing sets, based on a conjecture that large cluster projections contain small subsets that distinguish them.

For each distinguishing set S, we extend it to a cluster projection C as follows:

(1) Find the set $S_j \subseteq D_j$ of values connected to some $a_i \in S$.
(2) C contains S plus all $a_i \in D_i$ connected to some $a_j \in S_j$.

If $|C| > \delta$, we add C to C_i and prune any subsets of C from the remaining distinguishing sets $\mathcal{D}$. Additionally, any small distinguishing set S that could not extend into a larger projection is added to C_i.

The "SiblingStrength" and "ParticipationCount" terms account for overlapping projections. After computing projections from all attributes $A_j \neq A_i$, we intersect them to obtain the final set of cluster projections C_i on attribute A_i.

With the cluster projections C_i available, CACTUS next synthesizes full n-dimensional candidate clusters C in a level-wise manner.

We start with the 1-clusters $\mathcal{C}_1$ as the initial candidate clusters. Then for each level $k = 2 \ldots n$, we augment every $(k-1)$-cluster $c \in \mathcal{C}$ with each $c_k \in \mathcal{C}_k$ to form a candidate k-cluster c'. We add c' to the candidate set $\mathcal{C}'$ for level k if all attribute value pairs (a_i, a_k) across c and c_k are strongly connected (i.e. present in the inter-attribute summary σ_{ik}). This ensures the monotonicity property of clusters.

After considering all combinations of augmenting $(k-1)$-clusters to k-clusters, we update $\mathcal{C}$ to $\mathcal{C}'$ and proceed to level $k+1$. The final set $\mathcal{C}$ contains all candidate n-dimensional clusters over the full set of attributes.

The key properties that make this efficient are:

(1) Considering projections limits the search space.
(2) Monotonicity allows level-wise pruning of candidates.
(3) Intersections and unions are inexpensive set operations.

The final validation phase verifies whether each candidate cluster found in the previous phase satisfies the minimum support requirement on the original dataset D.

The validation procedure initializes counts $c(c)$ to zero for each candidate cluster $c \in \mathcal{C}$. It then scans the original dataset D once. For each tuple t, it increments the count $c(c)$ for every candidate cluster c whose region contains t.

After scanning D, for each cluster c we compute its expected support expectedSupport under an assumption of independence between attributes. This is calculated as α (support threshold parameter) multiplied by the product of relative cluster sizes $\frac{|c_i|}{|D_i|}$ across all attributes A_i.

Any cluster c whose actual support count $c(c)$ meets or exceeds the expected support is added to the final set of valid clusters $\mathcal{C}_v$ and returned.

The key advantages of this validation procedure are:

(1) It performs just a single scan over the original dataset D.
(2) For each tuple, it checks membership in multiple cluster regions simultaneously.
(3) It automatically prunes clusters that do not meet the minimum support criteria defined by the user parameter α.

8.8   Markov Clustering (MCL)

Markov Clustering (MCL) algorithm is one of the most prominent graph clustering algorithms, developed by Stijn van Dongen in the early 2000s. The MCL algorithm is based on the concept of random walks on graphs, exploiting the fact that random walks are more likely to remain within a cluster than to jump between clusters. By iteratively applying a series of mathematical operations on the adjacency matrix of the graph, the algorithm gradually isolates the clusters, making it a powerful tool for detecting communities in large, complex networks.

The MCL algorithm consists of three main steps: expansion, inflation, and pruning. Let's dive into the details of each step:

8.8.0.1 *Expansion*

The first step in the MCL algorithm is the expansion step. In this step, the adjacency matrix of the graph, denoted as $\mathbf{A}$, is raised to a power, $\mathbf{A}^r$, where r is a positive integer. This operation simulates random walks on the graph, as raising the adjacency matrix to a power corresponds to the probability of transitioning from one node to another after r steps.

8.8.0.2 *Inflation*

The second step is the inflation step. In this step, each element of the expanded adjacency matrix, $\mathbf{A}^r$, is raised to a power, $\mathbf{A}^{r \cdot \omega}$, where ω is the inflation factor. This operation amplifies the differences between strong and weak connections in the graph, effectively enhancing the contrast between potential clusters.

8.8.0.3 *Pruning*

The final step is the pruning step. In this step, the inflated matrix, $\mathbf{A}^{r \cdot \omega}$, is transformed by setting small values to zero, effectively removing weak connections from the graph. This step helps to isolate the clusters by focusing on the stronger, more significant connections within the graph.

The complete MCL algorithm can be summarized in the following pseudocode:

The key steps of the MCL algorithm can be explained as follows:

(1) The algorithm starts with the adjacency matrix $\mathbf{A}$ of the input graph and initializes the matrix $\mathbf{M}$ to be equal to $\mathbf{A}$.

(2) The algorithm then enters a loop that continues until the algorithm converges (i.e., the cluster assignments no longer change).

(3) In each iteration of the loop, the algorithm performs three main operations: A) Expansion: The matrix $\mathbf{M}$ is raised to the power r, which simulates r steps of random walks on the graph; B) Inflation: The elements

of the expanded matrix $\mathbf{M}^r$ are raised to the power ω, which amplifies the differences between strong and weak connections; C) Pruning: The inflated matrix $\mathbf{M}^{r\cdot\omega}$ is transformed by setting small values to zero, effectively removing weak connections.

(4) After the loop converges, the algorithm assigns each node to a cluster based on the final matrix $\mathbf{M}$.

The intuition behind the MCL algorithm is that random walks are more likely to stay within a cluster than to jump between clusters. By iteratively applying expansion, inflation, and pruning, the algorithm gradually isolates the clusters, allowing it to effectively detect communities in large, complex networks.

8.8.1   Running the MCL algorithm in R

To illustrate the application of the MCL algorithm, let's consider a simple example using the R programming language. We'll first create a sample graph and then apply the MCL algorithm to cluster the nodes.

Listing 8.9: Demonstrating MCL algorithm.

```
1  library(MCL)
2  ### Generate adjacency matrix of undirected graph
3  adjacency <-
4    matrix(c(0,1,1,1,0,0,0,0,0,1,0,1,1,1,0,0,0,0,1,1,
5        0,1,0,0,0,0,0,1,1,1,0,0,0,0,0,0,0,1,0,0,
6        0,1,1,0,0,0,0,0,0,1,0,1,0,0,0,0,0,0,1,1,
7        0,0,0,0,0,0,0,0,0,0,0,0,0,0,0,0,0,0,0,0,0),
8                    byrow=TRUE, nrow=9)
9  ### Plot graph (requires package igraph)
10 library(igraph)
11 gu <- graph.adjacency( adjacency, mode="undirected" )
12 plot( gu )
13
14 ### Run MCL
15 res<-mcl(x = adjacency, addLoops=TRUE, ESM = TRUE, inflation
        = 2, expansion = 2)
16 clusters <- res$Cluster
```

In this example, we first create a sample graph. We then apply the MCL algorithm using the mcl function from the MCL library, specifying the inflation and expansion factors as 2. The output of the mcl function is a list containing the cluster assignments for each node in the graph. We can extract the cluster assignments and use them for further analysis or visualization.

8.8.2 *Conclusions, advantages, limitations, and future research*

The Markov Clustering (MCL) algorithm is a powerful and versatile tool for graph clustering, with a wide range of applications in various domains. Its ability to detect communities in large, complex networks, coupled with its relatively robust performance in the presence of noise, makes it a valuable addition to the data scientist's toolbox.

Some of the key advantages of the MCL algorithm include:

(1) Effectiveness in detecting clusters in large, complex networks
(2) Robustness to noise and ability to handle weighted and directed graphs
(3) Simple and intuitive underlying concept based on random walks

However, the MCL algorithm also has some limitations:

(1) Sensitivity to the choice of parameters, such as the inflation and expansion factors
(2) Relatively high computational complexity, making it challenging to apply to extremely large graphs
(3) Potential for over-clustering or under-clustering, depending on the parameter settings

Despite these limitations, the MCL algorithm remains an active area of research, with ongoing efforts to improve its performance and extend its applicability. Some potential avenues for future research include:

(1) Developing adaptive parameter selection methods to improve the algorithm's robustness
(2) Exploring ways to reduce the computational complexity of the algorithm, enabling its application to even larger graphs
(3) Integrating the MCL algorithm with other clustering techniques or graph analysis methods to enhance its versatility and effectiveness

In conclusion, the Markov Clustering (MCL) algorithm is a powerful and versatile tool for graph clustering, with a solid mathematical foundation and a wide range of practical applications. By understanding its underlying principles and implementation details, data scientists can effectively leverage the MCL algorithm to uncover the hidden structures and communities within their complex, interconnected datasets (Figure 8.9).

Require: Adjacency matrix $\mathbf{A}$, expansion factor r, inflation factor ω
Ensure: Cluster assignment for each node

1: Initialize $\mathbf{M} = \mathbf{A}$
2: **while** not converged **do**
3:     $\mathbf{M} = \mathbf{M}^r$ Expansion
4:     $\mathbf{M} = \mathbf{M}^\omega$ Inflation
5:     $\mathbf{M} = \mathrm{Prune}(\mathbf{M})$
6: **end while**
7: Assign each node to a cluster based on the final matrix $\mathbf{M}$
8:
9: **return** cluster assignments for all nodes

Fig. 8.9: Markov clustering (MCL).

Chapter 9

Grid-Based Clustering Methods

Grid-based clustering methods are a category of clustering algorithms that quantize the data space into a finite number of cells or grids, and then perform clustering operations on the grid structures. The key idea is to reduce the vast data objects to a much smaller number of grid cells, which can make the clustering process more efficient, especially for large datasets. Grid-based clustering offers several advantages:

- **Efficiency:** By reducing the problem space to a finite number of cells, grid-based methods can significantly speed up the clustering process, especially for large datasets.
- **Scalability:** These methods can handle high-dimensional data more effectively by working with grids in subspaces.
- **Flexibility:** Grid-based methods can easily adapt to different data distributions and can be combined with other clustering techniques.

The general process for grid-based clustering involves the following steps:

(1) **Creating the grid structure:** The data space is divided into cells or grids of a particular shape and size. Common grid structures include rectangular cells, hexagonal cells, or trees.
(2) **Quantizing data to grids:** Each data object is mapped or quantized to a grid cell based on its location in the data space.
(3) **Clustering on grid cells:** With the data now in grid form, clustering operations are performed on the grid cells instead of the full data objects. Cells are classified as being part of a cluster or not part of a cluster.

(4) **Cluster formation:** The final clusters are formed by re-assembling the grid cells labeled as part of a cluster back into full data object groups.

There are several popular grid-based clustering algorithms:

(1) **STING (statistical information grid):** This algorithm constructs a hierarchical, multi-dimensional grid structure based on statistical information like means, maxima, minima, and variances stored in each grid cell. It uses successive layers to increase grid resolution in dense areas.

(2) **WaveCluster:** This grid-based algorithm leverages wavelet transforms to transform the spatial data into wavelet coefficients, which are mapped into a minimal-span multi-dimensional grid structure for clustering.

(3) **CLIQUE (clustering in QUEst):** CLIQUE identifies dense clusters in subspaces of maximum dimensionality using an Apriori property that says if a k-dimensional unit has a dense region, its (k-1)-dimensional projection also has a dense region.

The key advantages of grid-based methods include:

(1) Efficiency for clustering large datasets as only grid cells are processed.
(2) Ability to find clusters of arbitrary shapes.
(3) Minimal parameter inputs required.
(4) Some methods can handle noisy data well.

However, limitations include:

(1) The clustering quality is sensitive to the grid size and structure chosen.
(2) Memory usage may be high for high-dimensional data.
(3) Difficulties in finding natural/clear cluster boundaries.

Overall, grid-based methods offer a unique and often efficient approach to clustering large, complex datasets, trading off some resolution for computational scalability. Their usage is favored in domains like geographic information systems, astronomy, and scientific data analysis.

9.1 CLIQUE: Clustering in QUEst

CLIQUE (Clustering in QUEst) is a grid-based subspace clustering algorithm that automatically identifies dense clusters embedded in subspaces of high-dimensional data. It is particularly useful for clustering high-dimensional datasets, where traditional clustering algorithms may struggle to find meaningful clusters due to the "curse of dimensionality".

9.1.1 *How CLIQUE works*

The CLIQUE algorithm follows these key steps:

(1) **Partitioning the data space:** The first step is to partition the entire data space into non-overlapping rectangular units or cells of equal size.
(2) **Computing density at different dimensional subspaces:** For each dimensional subspace, CLIQUE identifies the dense cells by calculating the number of points within each cell.
(3) **Identifying dense regions:** A cell is marked as "dense" if the number of points inside it exceeds a user-specified threshold. This threshold can be different for different dimensional subspaces.
(4) **Cluster identification using the apriori property:** CLIQUE uses the Apriori property, which states that if a k-dimensional unit is dense, then its (k-1)-dimensional projection should also be dense. This property is used to find maximal regions of high density in all subspaces.
(5) **Generating minimal description for clusters:** The dense units in the highest dimensional subspace generate minimal clustering descriptions, representing the clusters.

The key strength of CLIQUE is its ability to handle high-dimensional data and automatically identify subspaces containing clusters. It can find clusters of arbitrary shape and size, and it is insensitive to the order of records in the input data.

The pseudocode for the CLIQUE algorithm is provided in Figure 9.1. First, the algorithm takes the multidimensional dataset D, the size of the grid units ξ, and the density threshold τ for each dimensional subspace as input.

Then, the data space is partitioned into n-dimensional grid units of size ξ. For each dimensional subspace k from 1 to n, the algorithm counts the number of points in each grid unit and marks the unit as dense if the count exceeds the density threshold τ_k. Then, using the Apriori property, the algorithm identifies the dense regions by checking if all $(k + 1)$-dimensional projections of a k-dimensional dense unit are dense. Then, for each n-dimensional dense unit, the algorithm generates a minimal cluster description representing the cluster. Finally, the set of minimal cluster descriptions C is returned as the output.

Require: D: Multidimensional dataset, ξ: Size of the grid units, τ: Density threshold for each dimensional subspace

Ensure: clusters

1: Partition D into n-dimensional grid units of size ξ
2: $dense\_units \leftarrow \emptyset$
3: **for** $k = 1$ to n **do**
4:     **for** each k-dimensional subspace S **do**
5:         **for** each unit u in S **do**
6:             Count the number of points in u
7:             **if** count $\geq \tau_k$ **then**
8:                 Mark u as dense in S
9:             **end if**
10:        **end for**
11:    **end for**
12: **end for**
13: **for** $k = n$ down to 1 **do**
14:    **for** each k-dimensional dense unit u **do**
15:        **if** all $(k + 1)$-dimensional projections of u are dense **then**
16:            $dense\_units$.add(u)
17:        **end if**
18:    **end for**
19: **end for**
20: $C \leftarrow \emptyset$
21: **for** each n-dimensional dense unit u **do**
22:    Generate a minimal cluster description for u
23:    C.add($minimal\_cluster\_description$)
24: **end for**
25:
26: **return** C

Fig. 9.1: CLIQUE algorithm.

9.1.2 *Using CLIQUE in R*

The subspace package provides a convenient interface for clustering high-dimensional data using subspace clustering algorithms like CLIQUE. Here is how you can use the subspace package to implement the CLIQUE algorithm:

```
Listing 9.1: Using CLIQUE algorithm
1  # Install and load the subspace package if not already
     installed
2  if (!require("subspace")) {
3    install.packages("subspace")
4  }
5  library(subspace)
6
7  # Generate sample data
8  set.seed(123)
9  data <- data.frame(x = rnorm(100), y = rnorm(100))
10
11 # Define the parameters
12 num_intervals <- 10
13 density_threshold <- 3
14
15 # Run the CLIQUE algorithm using the subspace package
16 result <- CLIQUE(data, xi = density_threshold, tau = 1/num_
     intervals)
17
18
19 #now plot the generated clustering
20 plot(result,data)
```

9.1.3 *Properties of the CLIQUE algorithm*

(1) **Subspace clustering:** CLIQUE can automatically identify and report clusters embedded in subspaces of the highest possible dimensionality.

(2) **Arbitrary shape and size clusters:** By using density-based clustering, CLIQUE can identify clusters of arbitrary shape and size.

(3) **Insensitivity to order of records:** The clustering results of CLIQUE are insensitive to the order in which records are presented.

(4) **Minimal description for clusters:** CLIQUE generates minimal descriptions for the identified clusters, making them easier to interpret and understand.

However, CLIQUE also has some limitations:

(1) It may not perform well on data with varying densities or clusters of diverse densities.
(2) The quality of the results depends on the grid size and density threshold parameters, which need to be carefully chosen.
(3) CLIQUE can be computationally expensive for very high-dimensional datasets due to the exponential number of possible subspaces.

Overall, CLIQUE is a powerful grid-based subspace clustering algorithm that can effectively handle high-dimensional data and automatically identify clusters in different subspaces. It has found applications in various domains, including bioinformatics, multimedia data analysis, and market basket analysis.

9.2 STING: Statistical Information Grid Clustering

STING (Statistical Information Grid) is a grid-based multi-resolution clustering technique that was developed for clustering spatial data objects. It constructs a hierarchical grid structure based on statistical information stored in each grid cell, allowing it to go into further detail and finer cells for high-density data regions.

9.2.1 *Overview of STING algorithm*

The STING algorithm involves the following key steps:

(1) **Partitioning into rectangular cells:** The spatial area is divided into rectangular cells at the highest level of the grid's hierarchical structure.
(2) **Calculating statistical metadata:** For each cell at every level, the statistical parameters like mean, maximum, minimum, and standard deviations are calculated for the contained objects.
(3) **Building the hierarchical grid structure:** The hierarchical grid structure is formed by subdividing the higher-level cells into further lower-level child cells based on specific criteria like the standard deviation or cell density.
(4) **Preparing the hierarchical tree:** A hierarchical tree structure is generated where higher-level cells point to their lower-level child cells.
(5) **Clustering based on levels:** The tree is traversed in a level-by-level fashion, marking higher-density regions as clusters of interest based on certain density thresholds.

(6) **Interactive selection:** The user can control the clustering granularity by selecting a higher or lower level in the hierarchical structure. Further analysis and querying can be done on the resulting clusters.

The pseudocode of the STING algorithm is provided in Figure 9.2. First, the spatial area is partitioned into rectangular cells at the highest level of the hierarchical grid structure. Then, for each cell at every level, statistical parameters like mean, maximum, minimum, and standard deviation are calculated for the data objects contained within the cell. Subsequently, the hierarchical grid structure is built by subdividing the higher-level cells into lower-level child cells based on certain criteria, such as the standard deviation or cell density. Then, a hierarchical tree structure is generated, where higher-level cells point to their lower-level child cells. The tree is traversed level by level, and higher-density regions are marked as clusters of interest based on certain density thresholds. Then, the user can interactively select the desired level of the hierarchical structure for clustering, allowing them to control the clustering granularity. Finally, the resulting clusters can be further analyzed and queried.

Note that this is a simplified pseudocode representation of the STING algorithm, and the actual implementation may involve additional details and optimizations.

1: Partition the spatial area into rectangular cells at the highest level
2: **for** each cell at every level **do**
3:    Calculate statistical parameters (mean, max, min, std. dev) for objects in the cell
4: **end for**
5: Build the hierarchical grid structure by subdividing cells based on criteria (e.g., std. dev, density)
6: Generate the hierarchical tree structure where higher-level cells point to lower-level child cells
7: Traverse the tree level by level
8: **for** each level **do**
9:    Mark higher-density regions as clusters based on density thresholds
10: **end for**
11: Allow user to select the desired level for clustering interactively
12: Analyze and query the resulting clusters

Fig. 9.2: STING algorithm.

The key strength of STING is its ability to store only statistical meta-data summaries in grid cells rather than the full details of all data objects, making it memory-efficient for large datasets. The hierarchical structure allows focusing on dense areas while ignoring sparse regions, making it computationally efficient.

9.2.2   *Properties of the STING algorithm*

(1) **Multi-resolution capabilities:** STING provides flexibility by allowing users to analyze data at different abstraction levels by browsing different levels of the hierarchical structure.
(2) **Discovery of clusters with arbitrary shape:** By using spatial data and density as heuristics, STING can identify clusters of arbitrary shape, size and density.
(3) **Parallelizability:** The hierarchical structure of STING can be processed in parallel for different branches to improve computational efficiency.
(4) **Semi-supervised clustering:** STING allows a semi-supervised approach where users can adjust the input parameters or browse the hierarchy to get the desired clustering results.

However, STING has some limitations:

(1) It relies on apriori chosen shape and size of cells at the lowest level, which can impact cluster discovery.
(2) The quality degrades for very high dimensional data due to the "curse of dimensionality".
(3) Clusters at arbitrary orientations may not be identified effectively.

Overall, STING provides an efficient grid-based technique for clustering large spatial datasets by using a hierarchical statistical grid structure. Its multi-resolution capabilities and ability to find arbitrary cluster shapes make it an important technique in spatial data mining and geographic information systems.

9.3   WaveCluster

WaveCluster uses wavelet transforms to convert the data into a different domain, where clusters can be more easily identified.

9.3.0.1 *WaveCluster algorithm steps*

The WaveCluster algorithm involves the following key steps:

(1) **Wavelet transform:** Apply a wavelet transform to the data to capture both frequency and location information.
(2) **Grid partitioning:** Partition the transformed data space into a grid of cells.
(3) **Cell density calculation:** Compute the density of each cell in the transformed space.
(4) **Cluster formation:** Identify and merge adjacent dense cells to form clusters.

9.3.0.2 *Advantages and limitations*

WaveCluster can efficiently identify clusters of arbitrary shapes and is robust to noise. However, it requires the computation of wavelet transforms, which can be computationally intensive.

9.4 GRIDCLUS

GRIDCLUS is another notable grid-based clustering algorithm that focuses on efficiently identifying clusters in large datasets by partitioning the data space into grids and merging dense cells. It extends the basic principles of grid-based clustering with specific strategies for handling high-dimensional data.

9.4.1 *Algorithm steps*

GRIDCLUS operates through the following steps:

(1) **Grid partitioning:** Similar to other grid-based methods, GRIDCLUS begins by partitioning the data space into a grid of cells called blocks. The size of the cells can be defined based on the data characteristics or user specifications.
(2) **Density calculation:** The density of each cell is calculated. Density can be defined in various ways, such as the number of data points within a cell or other statistical measures.
(3) **Dense cell identification:** Cells with densities above a predefined threshold are marked as dense. These dense cells are potential parts of clusters.

(4) **Cluster formation:** Dense cells that are adjacent are merged to form initial clusters. GRIDCLUS uses a specific neighborhood criterion to determine adjacency, which can be based on distance metrics or other proximity measures.

(5) **Cluster expansion and refinement:** Initial clusters are expanded by including adjacent dense cells. The algorithm also refines clusters by splitting or merging them based on density continuity and connectivity.

9.4.2 *Advantages and limitations*

GRIDCLUS algorithm has several known advantages:

- **Scalability:** GRIDCLUS is designed to handle large datasets efficiently by focusing on dense regions and ignoring sparse areas.
- **Flexibility:** The algorithm can adapt to various data distributions by adjusting grid sizes and density thresholds.
- **Noise robustness:** By concentrating on dense cells, GRIDCLUS can effectively ignore noise and outliers in the data.

At the same time, GRIDCLUS algorithm has several limitations, including:

- **Parameter sensitivity:** The performance of GRIDCLUS heavily depends on the choice of grid size and density thresholds. Improper settings can lead to poor clustering results.
- **Dimensionality:** While GRIDCLUS can handle high-dimensional data, its performance may degrade if the dimensionality becomes excessively high without appropriate dimensionality reduction techniques.

9.5 Applications of Grid-Based Clustering

Grid-based clustering methods have been successfully applied in various domains, including:

- **Image processing:** Identifying regions of interest in images based on pixel density.
- **Geographic data analysis:** Clustering spatial data points to identify regions with high activity or population density.
- **Bioinformatics:** Clustering gene expression data to find groups of co-expressed genes.

- **Market basket analysis:** Grouping items frequently purchased together in retail datasets.

9.6 Conclusion and Future Research

Grid-based clustering is a popular technique used in data mining and machine learning to partition a dataset into clusters. Unlike traditional clustering algorithms like k-means or hierarchical clustering, grid-based methods divide the data space into a finite number of cells (or grid units) and perform clustering operations on these cells rather than the individual data points. This approach can be particularly useful for handling large datasets and high-dimensional data, as it reduces the computational complexity and memory requirements of the clustering process.

While Grid-based clustering methods offer several advantages, including scalability and robustness to noise, they also face challenges such as:

- **Parameter selection:** Choosing appropriate parameters (e.g., grid size, density threshold) is crucial and can significantly impact the results.
- **Scalability:** While grid-based methods are generally efficient, handling extremely large datasets or very high-dimensional spaces can still be challenging.
- **Non-uniform data distributions:** These methods may struggle with data that has highly non-uniform distributions, as fixed grid sizes may not adapt well to varying data densities.

Future research directions for grid-based clustering include:

- **Adaptive grid methods:** Developing techniques to dynamically adjust grid sizes based on data density.
- **Parallel and distributed computing:** Leveraging modern computing architectures to improve the scalability of grid-based methods.
- **Integration with other methods:** Combining grid-based clustering with other clustering techniques to enhance robustness and accuracy.

Chapter 10

Deep Learning for Clustering

10.1 Introduction

Artificial Neural Networks (ANNs) have been widely used in various fields due to their ability to model complex relationships and patterns within data. This chapter explores the application of ANNs and deep learning in data clustering, a fundamental unsupervised learning task aimed at discovering the inherent structure within a dataset. We discuss how ANNs can be used for clustering, different architectures designed specifically for clustering tasks, and analyze their performance compared to traditional clustering algorithms. Additionally, we address the challenges and limitations associated with using ANNs for clustering and discuss potential future directions in this area.

10.2 Foundations of Artificial Neural Networks

Deep learning, a branch of machine learning, utilizes artificial neural networks (ANNs) to tackle intricate tasks. In recent years, deep learning has made substantial progress and has been employed in various fields such as computer vision, speech recognition, and natural language processing.

Artificial neural networks (ANNs) are computational models loosely inspired by the structure and function of the Biological brain or Biological Neural networks (BNNs). Like BNNs, ANNs consist of interconnected nodes or neurons that process information by applying mathematical operations to input data. However, while ANNs and BNNs are both neural networks that process information, they differ in their complexity, speed, and learning capability. While ANNs are simpler and faster, BNNs are incredibly complex and capable of carrying out a wide range of functions in the body. Moreover, BNNs are more capable of fault tolerance and are able to provide more robust performance in cases where there is partial damage to the network. Additionally, due to the fact that BNNs are composed of real neurons, they tend to be more energy-efficient, making them more suitable for applications that require large amounts of processing power.

The manifold hypothesis is an important concept in machine learning and data analysis, which suggests that high-dimensional data sets often have a low-dimensional structure, or manifold, that can be effectively described by a smaller number of dimensions. The idea is that many real-world data sets, such as images or speech signals, can be thought of as lying on a curved surface or manifold, which can be approximated by a set of lower-dimensional coordinates.

While it is true that the English language has a large vocabulary (700K words), most adult native speakers of the English language have a vocabulary that ranges from 20,000 to 35,000 words. Moreover, while many possible sentences can be constructed from a given set of words, not all of them are syntactically or semantically correct. This suggests that the underlying structure of language data may be lower-dimensional than the total number of possible word combinations, which is a basis for the manifold hypothesis. Machine learning algorithms can leverage this knowledge to extract the underlying structure of language data and model it more effectively.

The manifold hypothesis has been used to develop a variety of machine learning algorithms, such as principal component analysis (PCA), manifold learning, and deep learning, which are designed to extract the underlying structure of high-dimensional data sets and represent them in a more compact form. These algorithms have been used to solve a wide range of problems, such as image recognition, speech recognition, natural language processing, and predictive modeling, and have been shown to be highly effective in many applications. In that sense, machine learning algorithms such as ANNs aim to fit a flexible computing fabric to the manifold represented by the training examples instead of sewing pins.

When an ANN is trained on a set of input-output pairs, it learns to extract the underlying structure or manifold of the input data by adjusting the weights of its connections. This process can be thought of as finding a lower-dimensional representation of the input data that captures the most important features or variations of the data.

The learning process in ANNs can be viewed as a process of gradually refining the representation of the input data until it is sufficiently accurate for the desired task. By adjusting the weights of its connections in response to errors in its predictions, the ANN can iteratively learn to capture the important features and variations of the data, and approximate the underlying manifold. Moreover, the ability of ANNs to generalize to new, unseen data can also be explained by the manifold hypothesis. When an ANN has learned to capture the underlying manifold of the input data, it can make accurate predictions on new, unseen data points that lie on or near this manifold. This is because the ANN has learned to represent the input data in a more compact form, which allows it to generalize to new data points that share similar features or variations.

Over the past ten years, there have been many advancements in the field of training neural networks, which is considered to be more of an art than a science. Most of these advancements have been made through trial and error, where ideas and tricks have been added to a growing knowledge base about how to work with neural nets. When it comes to working with neural networks, there are three key factors to consider: choosing the right architecture for the task at hand, collecting data to train the neural net, and leveraging existing trained nets to improve the accuracy of new models. Interestingly, the same architecture can often be used for different tasks. Determining the amount of data required to train an ANN for a specific task is challenging to estimate, and it depends on several factors, such as the complexity of the task, the size of the neural network, and the desired level of accuracy. More complex networks with more parameters may require more training data to avoid overfitting. Generally, neural networks require a large amount of diverse training data to learn effectively. However, the required training set size can be reduced by using transfer learning, which involves incorporating knowledge from previously trained networks. Other factors that can affect the required training set size include the variability of the data and the level of noise in the data. Additionally, if the data is biased, meaning it does not accurately represent the overall population, the neural network may require more training data to generalize well. In many cases, ANNs can benefit from repetitive training rounds, during which they are shown the same examples repeatedly. Each round, also called an epoch, causes the ANN to be in a slightly different state, and reminding it of particular examples has been found to improve its ability to remember those examples. This strategy is considered a standard approach in training ANN.

Data augmentation can be used to artificially increase the size of the training dataset by creating additional, slightly modified versions of the existing data. This technique involves applying various transformations to the existing data. For example, if the data are images, we can rotate, crop, scale, flip, or add noise to the images. The idea behind data augmentation is that by providing the neural network with more diverse training examples, it will be able to generalize better and learn to recognize the underlying patterns in the data more effectively. Data augmentation can be especially useful in situations where the training dataset is limited, or where the cost of acquiring additional data is high.

Previously, it was believed that neural nets should be given as little responsibility as possible, and that data should be preprocessed before training. However, it is now believed that letting the neural net "discover" intermediate features and encodings by training on the "end-to-end problem" is more effective for human-like tasks. ANNs can be trained using supervised, unsupervised, or reinforcement learning to perform tasks such as classification, regression, and clustering. Deep learning is an ANN that consists of multiple layers that can learn complex features from high-dimensional data.

Before diving into the application of ANNs for clustering, it is essential to understand the basic components and functioning of ANNs. ANNs are computational models inspired by the human brain, consisting of interconnected nodes or neurons organized in layers. These neurons process information by receiving inputs, applying weights to these inputs, passing the weighted sum through an activation function, and producing an output.

10.2.1 *Network architecture*

ANNs can be categorized based on their architecture, including feedforward neural networks, recurrent neural networks, and convolutional neural networks, each suited for different types of data and tasks.

10.3 Deep Clustering: An Overview

Deep clustering refers to the use of deep learning models to facilitate the clustering process by learning representations that are more suitable for clustering tasks. The key idea is to leverage the feature learning ability of deep neural networks (DNNs) to transform raw data into a lower-dimensional space where clustering can be performed more effectively.

Traditional clustering methods often rely on predefined similarity measures like Euclidean distance, which may not be effective for high-dimensional or complex data. Deep clustering, on the other hand, uses neural networks to learn a similarity measure from the data itself, capturing intricate patterns and structures.

10.3.1 *Why deep clustering?*

There are several reasons why deep clustering has gained popularity:

- **High-dimensional data:** Traditional clustering algorithms often struggle with high-dimensional data due to the curse of dimensionality. Deep learning models can learn compact, informative representations that mitigate this issue.
- **Non-linear relationships:** Deep neural networks can capture complex, non-linear relationships in the data, which traditional linear methods cannot.
- **Scalability:** Deep learning models can scale to large datasets more effectively than many traditional clustering algorithms.

10.4 Types of Deep Clustering Methods

Deep clustering methods can be broadly categorized based on their network architectures and the way they integrate clustering objectives with representation learning. The main types are:

- Self-Organizing Maps (SOMs)
- Gaussian Mixture Models (GMMs) with Neural Networks
- Autoencoder-Based Clustering
- Clustering Deep Neural Networks (CDNN)
- Generative Adversarial Networks (GANs)
- Variational Autoencoders (VAEs)

10.4.1   *Self-organizing maps (SOMs)*

Self-Organizing Maps (SOM) are a type of artificial neural network introduced by Teuvo Kohonen in the 1980s. SOMs are particularly useful for clustering and visualizing high-dimensional data. Unlike other neural networks that are typically used for predictive tasks, SOMs are used for unsupervised learning, where the goal is to discover the inherent structure of the data.

SOMs can be seen as a clustering method that visualizes clusters in a more interpretable manner. SOMs achieve this by mapping high-dimensional input data into a lower-dimensional, typically two-dimensional, grid of neurons. Each neuron in the SOM corresponds to a prototype or reference vector. The training process involves updating these prototype vectors to reflect the structure of the input data, creating a topologically ordered map where similar data points are mapped to neighboring neurons.

The SOM algorithm can be broken down into several steps, which are repeated iteratively until convergence. Figure 10.1 given the pseudo-code for the SOM algorithm.

1: Initialize the weight vectors of the SOM neurons.
2: Set the initial learning rate and neighborhood radius.
3: **for** each epoch **do**
4:     **for** each input vector **x** in the dataset **do**
5:         Find the Best Matching Unit (BMU) for **x**.
6:         Update the weights of the BMU and its neighbors.
7:     **end for**
8:     Decrease the learning rate and neighborhood radius.
9: **end for**

Fig. 10.1: Self-organizing map (SOM) algorithm.

The algorithm begins with initializing the weight vectors of the neurons, often randomly. The learning rate and neighborhood radius are also initialized, typically to relatively high values. During each epoch, the algorithm processes each input vector **x** by finding the Best Matching Unit (BMU), which is the neuron whose weight vector is closest to **x**. Once the BMU is identified, its weight vector, along with those of its neighboring neurons, is adjusted to become more similar to **x**. This adjustment is done using a learning rate that decreases over time, ensuring that the adjustments become smaller as the algorithm converges. Similarly, the neighborhood radius, which determines the extent of the neighborhood around the BMU that gets updated, also decreases over time. This process is repeated for a predefined number of epochs or until the map stabilizes.

To run the SOM algorithm in R, we can use the "kohonen" package, which provides a comprehensive implementation of SOMs. Below is an example of how to use this package to train a SOM on a sample dataset.

In this example, we first install and load the "kohonen" package. We then load the Iris dataset and scale the data to ensure that all features have the same weight. The SOM grid is defined with a 5x5 hexagonal topology, and the SOM is trained using the "som" function. The training process is controlled by parameters such as "rlen" (the number of training iterations) and "alpha" (the learning rate). Finally, we plot the resulting SOM to visualize the trained map.

As we can see, Self-Organizing Maps offer a powerful tool for clustering and visualizing high-dimensional data. Their ability to create a topologically ordered representation of the data makes them particularly useful for exploring and understanding complex datasets.

Listing 10.1: Using SOM for clustering

```
1  # Install and load the kohonen package
2  install.packages("kohonen")
3  library(kohonen)
4
5  # Load a sample dataset
6  data(iris)
7  iris_data <- as.matrix(iris[, -5])
8
9  # Scale the data
10 iris_data <- scale(iris_data)
11
12 # Define the SOM grid
13 som_grid <- somgrid(xdim = 5, ydim = 5, topo = "hexagonal")
14
15 # Train the SOM
16 som_model <- som(iris_data, grid = som_grid, rlen = 100,
       alpha = c(0.05, 0.01))
17
18 # Plot the results
19 plot(som_model, type = "codes")
```

10.4.2   Gaussian mixture models (GMMs) with neural networks

GMMs are probabilistic models that assume data points are generated from a mixture of several Gaussian distributions. ANNs can be used to estimate the parameters of these Gaussian distributions, effectively performing clustering.

10.4.3   Autoencoder-based clustering

Autoencoders (AEs) are a type of neural network used to learn efficient codings of input data. An autoencoder consists of two parts: an encoder that maps the input data to a latent space, and a decoder that reconstructs the input data from the latent space. The goal is to minimize the reconstruction error.

In the context of clustering, the latent space learned by the autoencoder can be used for clustering. One of the pioneering works in this area is the Deep Embedded Clustering (DEC) algorithm, which simultaneously learns feature representations and cluster assignments by iteratively refining the cluster centers in the latent space.

$$L_{DEC} = L_{rec} + \alpha L_{clus} \qquad (10.1)$$

where L_{rec} is the reconstruction loss, L_{clus} is the clustering loss, and α is a weight parameter balancing the two losses.

The script in list 10.2 demonstrates how to use a neural network (autoencoder) for dimensionality reduction followed by a traditional clustering method (K-means) on the reduced data. This approach leverages the powerful feature extraction capabilities of neural networks for more effective clustering. The code begins with data Generation and normalization, where we generate synthetic data and normalize it. Then, we define an autoencoder with an encoding layer that reduces the dimensionality of the input data. Subsequently, the autoencoder is trained to reconstruct the input data.

Once the autoencoder is ready, we use the encoder part to transform the input data into a lower-dimensional representation. Then K-means clustering is performed on the encoded data and the clustering results are visualized using ggplot2.

Deep Learning for Clustering

245

Listing 10.2: Using auto-encoder for clustering

```r
1  # Load necessary libraries
2  library(keras)
3  library(stats)
4
5  # Generate synthetic data
6  data <- matrix(rnorm(1000), ncol=2)
7  labels <- sample(1:3, 500, replace=TRUE)
8  data[labels == 1, ] <- data[labels == 1, ] + 3
9  data[labels == 2, ] <- data[labels == 2, ] - 3
10
11 data <- scale(data)
12
13 # Define autoencoder architecture
14 input_dim <- ncol(data)
15 encoding_dim <- 2   # Dimension of encoded representation
16
17 input_layer <- layer_input(shape = c(input_dim))
18 encoded <- input_layer %>%
19   layer_dense(units = 8, activation = 'relu') %>%
20   layer_dense(units = encoding_dim, activation = 'relu')
21
22 decoded <- encoded %>%
23   layer_dense(units = 8, activation = 'relu') %>%
24   layer_dense(units = input_dim, activation = 'sigmoid')
25
26 autoencoder <- keras_model(input_layer, decoded)
27
28 # Compile the autoencoder
29 autoencoder %>% compile(optimizer = 'adam', loss = 'mean_squared_
      error')
30
31 # Train the autoencoder
32 history <- autoencoder %>% fit(data, data,epochs = 50,
33   batch_size = 32, shuffle = TRUE, validation_split = 0.2)
34
35 # Extract the encoder part of the autoencoder
36 encoder <- keras_model(input_layer, encoded)
37
38 # Encode the data to the latent space
39 encoded_data <- encoder %>% predict(data)
40
41 # Perform K-means clustering on the encoded data
42 kmeans_result <- kmeans(encoded_data, centers = 3, nstart = 20)
43
44 # Visualize the clustering result
45 library(ggplot2)
46 encoded_df <- as.data.frame(encoded_data)
47 encoded_df$cluster <- as.factor(kmeans_result$cluster)
48
49 ggplot(encoded_df, aes(x = V1, y = V2, color = cluster)) +
      geom_point() +
50   labs(title = "Clustering Result Using Neural Network",
51       x = "Encoded Feature 1", y = "Encoded Feature 2") +
            theme_minimal()
```

10.4.3.1 *Deep embedded clustering (DEC)*

The DEC algorithm is a widely recognized method that integrates autoencoders with clustering. It aims to refine cluster assignments iteratively while learning a feature representation suitable for clustering.

- **Pretraining:** Train the autoencoder to minimize reconstruction loss on the input data.
- **Clustering initialization:** Initialize cluster centers in the latent space using a traditional clustering method like k-means.
- **Iterative refinement:** Alternate between updating cluster assignments and refining the latent space representations.

The loss function for DEC can be expressed as:

$$L_{DEC} = KL(P||Q) \tag{10.2}$$

where KL denotes the Kullback-Leibler divergence, P is the target distribution, and Q is the distribution of soft cluster assignments.

10.4.3.2 *Algorithm implementation*

Here is a basic outline of the DEC algorithm:

Algorithm 10.1 Deep Embedded Clustering (DEC)

1: **Input:** Data X, number of clusters K, weight parameter α
2: **Output:** Cluster assignments
3: Train autoencoder to minimize reconstruction loss L_{rec}
4: Initialize cluster centers using k-means on latent space
5: **repeat**
6:     Compute soft cluster assignments
7:     Update target distribution
8:     Update autoencoder and cluster centers to minimize L_{DEC}
9: **until** convergence
10: **return** Cluster assignments

10.4.4 *Clustering deep neural networks (CDNN)*

Clustering Deep Neural Networks (CDNN) are designed to optimize a clustering-specific objective directly during the training of the network. These networks are typically pre-trained on large datasets to learn useful features and then fine-tuned using clustering objectives. The clustering loss guides the learning process to form compact and well-separated clusters in the feature space.

10.4.4.1 *Algorithm implementation*

A typical CDNN framework might involve the following steps:

- **Pretraining:** Use a deep neural network to learn a good feature representation.
- **Clustering objective:** Fine-tune the network using a clustering-specific loss function.

The loss function for a CDNN could be:

$$L_{CDNN} = L_{pretrain} + \beta L_{clus} \qquad (10.3)$$

where $L_{pretrain}$ is the pretraining loss, L_{clus} is the clustering loss, and β is a weight parameter.

10.4.5 *Generative adversarial networks (GANs)*

Generative Adversarial Networks (GANs) consist of a generator and a discriminator network that compete against each other. The generator tries to create realistic data samples, while the discriminator tries to distinguish between real and generated samples. This adversarial process can generate high-quality data samples.

In deep clustering, GANs can be used to learn the distribution of the data and generate synthetic samples that enhance the clustering process. The ClusterGAN framework integrates clustering into the GAN structure, allowing it to generate data points that belong to specific clusters.

10.4.5.1 *ClusterGAN framework*

The ClusterGAN framework involves the following steps:

- **Generator:** Generates data samples from a latent space.
- **Discriminator:** Distinguishes between real and generated samples.
- **Clustering objective:** The latent space is structured to facilitate clustering.

The loss functions for the generator and discriminator are:

$$L_G = -\mathbb{E}_{z \sim p_z}[\log D(G(z))] \tag{10.4}$$

$$L_D = -\mathbb{E}_{x \sim p_{data}}[\log D(x)] - \mathbb{E}_{z \sim p_z}[\log(1 - D(G(z)))] \tag{10.5}$$

10.4.5.2 *Algorithm implementation*

A basic implementation of ClusterGAN involves:

Algorithm 10.2 ClusterGAN

1: **Input:** Data X, latent space Z
2: **Output:** Clustered latent representations
3: **repeat**
4:     Sample random noise z from Z
5:     Generate data samples using $G(z)$
6:     Train D to distinguish real and fake samples
7:     Train G to fool D
8: **until** convergence
9: **return** Clustered latent representations

10.4.6   *Variational autoencoders (VAEs)*

Variational Autoencoders (VAEs) are a type of generative model that learns a probabilistic mapping from data to a latent space. VAEs introduce a regularization term to the latent space, encouraging the learned representations to follow a predefined distribution (typically Gaussian).

In clustering, VAEs can generate new data samples from the latent space, aiding in the identification of clusters. The Variational Deep Embedding (VaDE) framework uses a VAE to learn a mixture of Gaussians in the latent space, facilitating clustering by associating each data point with a Gaussian component.

10.4.6.1 *Variational deep embedding (VaDE)*

VaDE combines the VAE framework with Gaussian Mixture Models (GMMs) in the latent space. The key idea is to model the latent space as a mixture of Gaussian distributions, each corresponding to a cluster.

The VAE loss function consists of two parts:

$$L_{VAE} = L_{rec} + L_{KL} \qquad (10.6)$$

where L_{rec} is the reconstruction loss and L_{KL} is the Kullback-Leibler divergence between the learned latent distribution and the prior.

10.4.6.2 *Algorithm implementation*

An implementation of VaDE involves:

Algorithm 10.3 Variational Deep Embedding (VaDE)

1: **Input:** Data X, number of clusters K
2: **Output:** Cluster assignments
3: Initialize VAE with random parameters
4: **repeat**
5:   Sample from the posterior distribution $q(z|x)$
6:   Compute reconstruction loss L_{rec}
7:   Compute KL divergence L_{KL}
8:   Update VAE parameters to minimize L_{VAE}
9: **until** convergence
10: Fit GMM to the latent space
11: **return** Cluster assignments from GMM

10.5 Applications of Deep Clustering

Deep clustering has been applied to various domains, demonstrating its versatility and effectiveness:

10.5.1 *Image clustering*

Deep clustering methods have been successfully used for clustering image data, benefiting from the powerful feature extraction capabilities of convolutional neural networks (CNNs). Techniques like Deep Convolutional Embedded Clustering (DCEC) extend DEC by incorporating convolutional layers, allowing the model to handle spatial hierarchies in images.

For example, the MNIST dataset, consisting of handwritten digit images, is a popular benchmark for clustering methods. Deep clustering approaches have shown remarkable performance in separating different digit classes, significantly outperforming traditional methods.

10.5.2   *Text clustering*

Recurrent neural networks (RNNs) and transformer models have been employed to cluster textual data, capturing the sequential nature and contextual information of texts. Methods like the Deep Embedded Topic Model (DETM) integrate topic modeling with deep learning, enabling effective clustering of documents.

Clustering news articles into topics is a common application of text clustering. Deep clustering methods can uncover latent topics and group similar articles together, providing a better understanding of the underlying themes.

10.5.3   *Speech and audio processing*

Deep clustering has been applied to tasks like speech separation and audio event detection, leveraging the ability to learn representations that separate different sound sources. The Deep Clustering (DPCL) method uses a deep embedding network to assign time-frequency bins to different speakers, improving speech separation performance.

The cocktail party problem, where multiple speakers talk simultaneously, poses a significant challenge for traditional methods. Deep clustering approaches have shown promise in isolating individual speakers from mixed audio signals, demonstrating their effectiveness in real-world scenarios.

10.6   Conclusions, Challenges and Future Directions

Deep clustering represents a powerful fusion of deep learning and traditional clustering techniques, offering significant improvements in handling high-dimensional, complex data. By learning data representations that are inherently more suitable for clustering, deep clustering methods have opened new avenues in data analysis and pattern recognition. As research continues to address current challenges and explore new directions, deep clustering is poised to become an indispensable tool in the data scientist's arsenal.

10.6.1   *Challenges*

Despite the significant progress, deep clustering faces several challenges:

10.6.1.1 *Scalability*

Training deep clustering models on very large datasets remains computationally expensive. Efficient training algorithms and scalable architectures are needed to address this issue.

10.6.1.2 *Interpretability*

The learned representations and clustering results are often difficult to interpret, which can be a barrier for certain applications. Developing techniques to enhance the interpretability of deep clustering models is an important area of research.

10.6.1.3 *Theoretical understanding*

There is a need for a deeper theoretical understanding of why and how deep clustering methods work, which can guide the development of more effective algorithms. Establishing theoretical foundations will help in designing models with better performance guarantees.

10.6.2 *Future research directions*

Future research directions in deep clustering include:

10.6.2.1 *Theoretical exploration*

Developing theoretical frameworks to understand and improve deep clustering methods. This includes investigating the convergence properties, stability, and robustness of deep clustering algorithms.

10.6.2.2 *New architectures*

Exploring new network architectures, such as graph neural networks, for clustering complex data types. Graph-based models can capture the relationships between data points more effectively, leading to improved clustering performance.

10.6.2.3 *Domain adaptation*

Enhancing the ability of deep clustering methods to generalize across different domains and tasks. Transfer learning and domain adaptation techniques can be incorporated to adapt deep clustering models to new datasets with minimal retraining.

10.6.2.4 *Semi-supervised and unsupervised learning*

Investigating semi-supervised and unsupervised learning approaches for deep clustering. Incorporating limited labeled data can improve clustering performance, while fully unsupervised methods can handle scenarios where no labeled data is available.

Chapter 11

Spectral Clustering

11.1 Background and Motivation

Spectral clustering has its roots in graph theory and spectral graph partitioning, which offer a fundamentally different approach to clustering compared to traditional methods. The central idea behind spectral clustering is to construct a graph from the data, where each node represents a data point, and the edges represent the similarity between points. This graph-based representation allows the relationships among data points to be captured in a structured way. By analyzing the eigenvectors of matrices associated with this graph, specifically the Laplacian matrix, we can identify clusters in the data. Traditional clustering methods, such as K-means and hierarchical clustering, have significant limitations. K-means, for instance, assumes that clusters are spherical and of equal size. This assumption often causes K-means to struggle with data that have complex, non-convex shapes or clusters of varying sizes. Hierarchical clustering, on the other hand, builds a hierarchy of clusters by either merging smaller clusters into larger ones or by splitting larger clusters into smaller ones. While hierarchical clustering can reveal the structure of data at different levels of granularity, it is highly sensitive to noise and outliers. Moreover, hierarchical clustering can become computationally expensive when dealing with large datasets, making it less practical for big data applications. Spectral clustering addresses many of these limitations by taking advantage of the properties of eigenvectors derived from the similarity matrix. This allows it to capture complex cluster shapes that traditional methods cannot handle effectively. For example, clusters that are non-convex or have irregular boundaries can be more accurately identified using spectral clustering. Additionally, spectral clustering is robust to variations in cluster sizes and densities, which is often

253

a challenge for K-means. By leveraging the eigenvectors of a similarity matrix, spectral clustering can also reduce the impact of noise and outliers,
leading to more stable and accurate clustering results. In spectral clustering, the first step is to construct a similarity matrix from the data. This
matrix is typically derived from pairwise similarities between data points,
where higher values indicate greater similarity. One common approach is to
use a Gaussian (RBF) kernel to compute these similarities. Once the similarity matrix is constructed, the degree matrix is formed by summing the
similarities for each data point. The Laplacian matrix, which is the difference between the degree matrix and the similarity matrix, encapsulates the
structure of the data in a way that is suitable for spectral analysis. The next
step involves computing the eigenvectors of the Laplacian matrix. These
eigenvectors provide a new representation of the data, where the geometric properties of the original data are transformed into a lower-dimensional
space that reveals the underlying cluster structure. By selecting the eigenvectors corresponding to the smallest eigenvalues, we can project the data
into this lower-dimensional space. The resulting representation allows us to
apply a standard clustering algorithm, such as K-means, to identify clusters. This approach enables spectral clustering to be particularly effective
in situations where traditional methods fail. For instance, in image segmentation, where the goal is to partition an image into meaningful regions,
spectral clustering can accurately capture regions with complex shapes and
varying textures. In social network analysis, spectral clustering can identify
communities within networks, even when these communities are not easily separable by traditional methods. In bioinformatics, spectral clustering
has been used to group gene expression data, revealing insights into gene
function and regulation.

11.2   Graph Theory Basics

Understanding spectral clustering requires some basic knowledge of graph
theory. In this section, we will introduce fundamental concepts such as
graphs, adjacency matrices, and Laplacian matrices, which are essential for
comprehending the spectral clustering algorithm.

11.2.1   *Graphs and adjacency matrices*

A graph $G = (V, E)$ consists of a set of vertices V and a set of edges E that
connect pairs of vertices. In the context of spectral clustering, each vertex

represents a data point, and each edge represents the similarity between a pair of data points. The graph can be represented using an adjacency matrix A, which is a square matrix where the element A_{ij} denotes the similarity between the i-th and j-th vertices.

Formally, the adjacency matrix A for an undirected graph with n vertices is defined as:

$$A_{ij} = \begin{cases} s(x_i, x_j) & \text{if there is an edge between vertices } i \text{ and } j, \\ 0 & \text{otherwise,} \end{cases}$$

where $s(x_i, x_j)$ is a similarity measure between the data points x_i and x_j. Common choices for the similarity measure include the Gaussian (RBF) kernel, defined as:

$$s(x_i, x_j) = \exp\left(-\frac{\|x_i - x_j\|^2}{2\sigma^2}\right),$$

and the k-nearest neighbors approach, where $A_{ij} = 1$ if x_i is among the k-nearest neighbors of x_j, and 0 otherwise.

11.2.2 Degree matrix

The degree matrix D is a diagonal matrix that contains information about the degrees of the vertices. The degree of a vertex is the sum of the weights of the edges connected to it. For an adjacency matrix A, the degree matrix D is defined as:

$$D_{ii} = \sum_{j=1}^{n} A_{ij},$$

and $D_{ij} = 0$ for $i \neq j$.

11.2.3 Laplacian matrix

The Laplacian matrix L of a graph is a key component in spectral clustering. It captures the structure of the graph and is defined as:

$$L = D - A,$$

where D is the degree matrix and A is the adjacency matrix. The Laplacian matrix plays a central role in spectral graph theory and clustering, as its eigenvectors are used to embed the graph vertices in a lower-dimensional space.

In addition to the unnormalized Laplacian matrix L, there are normalized versions that are often used in spectral clustering. The normalized Laplacian matrix L_{sym} is defined as:

$$L_{sym} = D^{-1/2} L D^{-1/2} = I - D^{-1/2} A D^{-1/2},$$

where I is the identity matrix. Another normalized version is the random walk normalized Laplacian L_{rw}, defined as:

$$L_{rw} = D^{-1} L = I - D^{-1} A.$$

11.2.4   *Properties of the Laplacian matrix*

The Laplacian matrix has several important properties that make it useful for clustering:

- L is symmetric and positive semi-definite.
- The smallest eigenvalue of L is 0, with the corresponding eigenvector being the constant vector $\mathbf{1}$.
- The multiplicity of the eigenvalue 0 indicates the number of connected components in the graph.

11.2.5   *Spectral embedding*

Spectral embedding involves using the eigenvectors of the Laplacian matrix to project the data points into a lower-dimensional space. This process reveals the structure of the data and helps identify clusters. Specifically, for spectral clustering, we compute the first k eigenvectors of the Laplacian matrix L (or L_{sym} or L_{rw}), corresponding to the smallest eigenvalues, and use these eigenvectors to form a new representation of the data. Each data point is then represented by its components in these eigenvectors, effectively embedding the data into a k-dimensional space.

In the next sections, we will delve deeper into the spectral clustering algorithm, exploring how these graph-theoretical concepts are utilized to achieve effective clustering results.

11.3   Spectral Clustering Algorithm

The spectral clustering algorithm provides a robust and versatile method for partitioning data into clusters based on the eigenvectors of matrices derived from the data's similarity graph. In this section, we will detail each step of the spectral clustering process, from forming the similarity matrix to assigning the original data points to clusters.

11.3.1 Eigenvector decomposition

The core of spectral clustering involves the eigenvector decomposition of the Laplacian matrix L. We compute the first k eigenvectors corresponding to the smallest eigenvalues of L. These eigenvectors form a matrix $U \in \mathbb{R}^{n \times k}$, where each row of U represents a data point in the reduced k-dimensional space.

For normalized spectral clustering, we compute the eigenvectors of L_{sym}. These eigenvectors are orthogonal and provide a basis that captures the intrinsic geometric structure of the data.

11.3.2 Clustering in reduced space

The rows of the matrix U obtained from the eigenvector decomposition represent the data points in the new k-dimensional space. To prepare for clustering, we often normalize these rows to have unit length:

$$Y_{ij} = \frac{U_{ij}}{\sqrt{\sum_j U_{ij}^2}}.$$

This normalization step ensures that the clustering algorithm operates on data points that lie on a unit sphere, improving the robustness of the clustering results.

We then apply a standard clustering algorithm, such as K-means, to the rows of the normalized matrix Y. The choice of k, the number of clusters, is critical and is often determined through cross-validation or domain knowledge.

11.3.3 Assigning original points

The final step in spectral clustering is to assign the original data points to the identified clusters. Each row of the matrix U corresponds to a data point in the original dataset, and the clustering algorithm applied in the reduced space partitions these points into clusters. The cluster assignment from the reduced space is directly transferred to the original data points, completing the spectral clustering process.

11.4 Analysis and Interpretation

11.4.1 Ideal case

In an ideal scenario where clusters are well-separated, spectral clustering can perfectly recover the clusters. The eigenvectors of the Laplacian matrix will clearly distinguish between different clusters, and the clustering algorithm in the reduced space will accurately reflect these separations.

11.4.2   *Practical considerations*

In practice, the performance of spectral clustering depends on several factors:

- **Choice of similarity matrix:** The method used to construct the similarity matrix A can significantly affect the results. Careful selection of parameters, such as σ in the Gaussian kernel or k in the k-nearest neighbors method, is crucial.
- **Number of clusters** k**:** Choosing the correct number of clusters is essential for accurate results. This can often be done using cross-validation, the elbow method, or domain-specific knowledge.
- **Normalization:** Using the normalized Laplacian often improves performance by ensuring that the data points lie on a unit sphere, which can help the clustering algorithm to produce better results.

Spectral clustering's ability to handle complex cluster shapes, varying densities, and its robustness to noise and outliers make it a powerful tool in the machine learning toolkit. However, like any method, it requires careful implementation and parameter tuning to achieve optimal results.

11.5   Multiscale Spectral Clustering

Spectral clustering can be extended to handle data at multiple scales, improving robustness and performance on complex datasets. This extension, known as multiscale spectral clustering, leverages the ability to analyze data at different levels of granularity, capturing both local and global structures. In this section, we will discuss the motivation behind multiscale spectral clustering, the methods used to implement it, and its advantages in handling complex datasets.

11.5.1   *Motivation for multiscale analysis*

In many real-world datasets, clusters can exist at different scales. For example, in image segmentation, a picture may contain both small, detailed regions and larger, more general areas. Similarly, in social networks, tightly-knit communities may be nested within larger groups. Traditional spectral clustering, while powerful, typically operates at a single scale, potentially missing finer or broader structures present in the data.

Multiscale spectral clustering addresses this limitation by integrating information from multiple scales. This approach enhances the robustness of clustering results and improves performance on datasets with hierarchical or nested cluster structures.

11.5.2 Constructing multiscale similarity matrices

The key to multiscale spectral clustering lies in constructing multiple similarity matrices, each representing the data at a different scale. One common approach is to vary the parameters used in the similarity computation, such as the bandwidth σ in the Gaussian (RBF) kernel. By creating similarity matrices with different σ values, we can capture relationships among data points at various levels of detail.

Formally, let $\sigma_1, \sigma_2, \ldots, \sigma_m$ represent a sequence of bandwidth parameters. For each σ_i, we construct a similarity matrix $A^{(\sigma_i)}$ as follows:

$$A_{jk}^{(\sigma_i)} = \exp\left(-\frac{\|x_j - x_k\|^2}{2\sigma_i^2}\right).$$

These matrices capture the similarity between data points at different scales, ranging from fine (small σ) to coarse (large σ).

11.5.3 Combining multiscale information

Once we have multiple similarity matrices, the next step is to combine them into a single multiscale similarity matrix. This can be done through various methods, such as averaging or weighted summation. A simple and effective approach is to compute the average of the similarity matrices:

$$A_{jk} = \frac{1}{m} \sum_{i=1}^{m} A_{jk}^{(\sigma_i)}.$$

This combined similarity matrix A integrates information from all scales, providing a comprehensive view of the relationships between data points.

11.5.4 Multiscale Laplacian matrix

With the combined similarity matrix A, we proceed to construct the degree matrix D and the Laplacian matrix L as in traditional spectral clustering:

$$D_{jj} = \sum_{k=1}^{n} A_{jk},$$

$$L = D - A.$$

For normalized multiscale spectral clustering, we use the normalized Laplacian matrix L_{sym}:

$$L_{sym} = D^{-1/2} L D^{-1/2} = I - D^{-1/2} A D^{-1/2}.$$

11.5.5   *Eigenvector decomposition and clustering*

As with traditional spectral clustering, the eigenvector decomposition of the multiscale Laplacian matrix L or L_{sym} is performed. The first k eigenvectors corresponding to the smallest eigenvalues are computed, forming a matrix $U \in \mathbb{R}^{n \times k}$.

The rows of U represent the data points in the reduced k-dimensional space. These rows are then normalized to unit length:

$$Y_{jk} = \frac{U_{jk}}{\sqrt{\sum_k U_{jk}^2}}.$$

Clustering is performed on the rows of the normalized matrix Y using a standard algorithm such as K-means, resulting in the assignment of data points to clusters.

11.5.6   *Advantages of multiscale spectral clustering*

Multiscale spectral clustering offers several advantages over traditional single-scale methods:

- **Improved robustness:** By incorporating information from multiple scales, the method becomes less sensitive to the choice of a single scale parameter, reducing the risk of poor clustering results due to an inappropriate scale.
- **Enhanced performance:** The integration of local and global structures allows the algorithm to better capture complex cluster shapes and nested structures within the data.
- **Versatility:** Multiscale spectral clustering can be applied to a wide range of datasets, including those with hierarchical or multi-level cluster structures, making it a versatile tool for various applications.

In conclusion, multiscale spectral clustering extends the capabilities of traditional spectral clustering by leveraging the power of multiple scales. This approach enhances robustness and performance, making it particularly effective for complex datasets with intricate cluster structures. The following sections will explore specific applications of multiscale spectral clustering and provide examples of its implementation in different domains.

11.6 Sparse Spectral Clustering

For large datasets, spectral clustering can become computationally expensive due to the need to compute and store the full similarity matrix, which has a size of $O(n^2)$. Sparse spectral clustering methods address this issue by using sparsified versions of the similarity matrix, significantly reducing computational complexity and memory requirements while preserving the essential structure of the data. In this section, we will discuss the principles behind sparse spectral clustering, techniques for sparsifying the similarity matrix, and the benefits and challenges associated with this approach.

11.6.1 *Principles of sparse spectral clustering*

The main idea behind sparse spectral clustering is to approximate the full similarity matrix A with a sparse matrix $\tilde{A}$, where most of the elements are zero. This sparsification retains the most significant similarities between data points while discarding weaker connections that have less impact on the clustering results. By operating on this sparsified matrix, we can achieve substantial computational savings.

Sparse spectral clustering leverages the fact that in many datasets, meaningful clusters can be identified by focusing on the strongest pairwise similarities. This approach is particularly effective in high-dimensional data, where many similarities are close to zero and contribute little to the overall cluster structure.

11.6.2 *Techniques for sparsifying the similarity matrix*

There are several techniques for creating a sparse approximation of the similarity matrix. The most common methods include the k-nearest neighbors approach and the ϵ-neighborhood method.

11.6.2.1 *k-Nearest neighbors*

In the k-nearest neighbors (k-NN) approach, we construct the sparse similarity matrix $\tilde{A}$ by retaining only the k largest similarities for each data point. Formally, for each data point x_i, we keep the similarities to its k

nearest neighbors and set the rest to zero:

$$\tilde{A}_{ij} = \begin{cases} A_{ij} & \text{if } x_j \text{ is among the } k \text{ nearest neighbors of } x_i, \\ 0 & \text{otherwise.} \end{cases}$$

This results in a sparse matrix where each row has at most k non-zero entries. The choice of k is crucial and typically depends on the specific dataset and the desired balance between sparsity and fidelity to the original similarity structure.

11.6.2.2 ϵ-Neighborhood

In the ϵ-neighborhood method, we retain only the similarities that exceed a certain threshold ϵ. The sparse similarity matrix $\tilde{A}$ is defined as:

$$\tilde{A}_{ij} = \begin{cases} A_{ij} & \text{if } A_{ij} > \epsilon, \\ 0 & \text{otherwise.} \end{cases}$$

This approach effectively filters out weak similarities, keeping only those that indicate strong connections between data points. The threshold ϵ is chosen based on the distribution of similarity values in the dataset.

11.6.2.3 *Combination methods*

In practice, it is often beneficial to combine the k-nearest neighbors and ϵ-neighborhood methods. For instance, we can construct a sparse matrix by retaining the k largest similarities for each data point, but only if they exceed a certain threshold ϵ. This combined approach can provide a more refined sparsification, capturing both local neighborhood relationships and strong global similarities.

11.6.3 *Computing the sparse Laplacian*

Once the sparse similarity matrix $\tilde{A}$ is constructed, the next steps follow the standard spectral clustering procedure. We compute the degree matrix $\tilde{D}$ from $\tilde{A}$ as:

$$\tilde{D}_{ii} = \sum_{j=1}^{n} \tilde{A}_{ij},$$

and then the sparse Laplacian matrix $\tilde{L}$ is given by:

$$\tilde{L} = \tilde{D} - \tilde{A}.$$

For normalized sparse spectral clustering, we use the normalized sparse Laplacian matrix $\tilde{L}_{sym}$:

$$\tilde{L}_{sym} = \tilde{D}^{-1/2}\tilde{L}\tilde{D}^{-1/2} = I - \tilde{D}^{-1/2}\tilde{A}\tilde{D}^{-1/2}.$$

11.6.4 *Eigenvector decomposition and clustering*

With the sparse Laplacian matrix $\tilde{L}$ or $\tilde{L}_{sym}$, we perform eigenvector decomposition to obtain the first k eigenvectors corresponding to the smallest eigenvalues. These eigenvectors form a matrix $\tilde{U} \in \mathbb{R}^{n \times k}$, where each row represents a data point in the reduced k-dimensional space.

The rows of $\tilde{U}$ are then normalized to unit length:

$$\tilde{Y}_{ij} = \frac{\tilde{U}_{ij}}{\sqrt{\sum_j \tilde{U}_{ij}^2}}.$$

A standard clustering algorithm, such as K-means, is applied to the rows of the normalized matrix $\tilde{Y}$ to partition the data points into clusters.

11.6.5 *Advantages and challenges of sparse spectral clustering*

11.6.5.1 *Advantages*

- **Reduced computational complexity:** By sparsifying the similarity matrix, the computational cost of eigenvector decomposition is significantly reduced, making spectral clustering feasible for large datasets.
- **Memory efficiency:** Sparse matrices require less storage space, allowing the algorithm to handle larger datasets within the memory constraints of typical computing environments.
- **Preservation of local structures:** The sparsification process focuses on retaining the most significant similarities, often corresponding to local neighborhood structures, which are crucial for accurate clustering.

11.6.5.2 *Challenges*

- **Choice of parameters:** Selecting appropriate values for k and ϵ is critical for effective sparsification. Poor choices can lead to the loss of important structural information or insufficient sparsity.
- **Potential loss of global structure:** While sparsification preserves local structures, there is a risk of losing global structural information, which can impact the overall clustering quality.

- **Implementation complexity:** Implementing sparse spectral clustering requires careful design and optimization to balance sparsity and fidelity, which can add to the complexity of the algorithm.

In conclusion, sparse spectral clustering extends the applicability of spectral methods to large datasets by reducing computational and memory demands. By carefully constructing sparse similarity matrices, this approach retains essential structural information, enabling efficient and effective clustering. The following sections will explore specific applications and case studies demonstrating the power and versatility of sparse spectral clustering.

11.7   The Normalized Cuts Algorithm

Spectral clustering is a powerful technique for identifying clusters in data based on the eigenvectors of matrices derived from the data's similarity graph. One of the most influential spectral clustering algorithms is the Normalized Cuts (NCuts) algorithm. The NCuts algorithm, introduced by Shi and Malik (2000), addresses the challenge of segmenting a graph into disjoint, yet balanced, clusters. Unlike traditional methods that focus solely on the minimization of edge cuts, NCuts simultaneously considers the association within clusters and the separation between them, resulting in more meaningful partitions.

The key idea behind NCuts is to represent the data as a graph, where nodes correspond to data points, and edges reflect the similarity between pairs of points. The objective is to partition the graph into two sets such that the normalized cut, which measures the disconnection between the sets relative to the connection within them, is minimized. This method ensures that the resulting clusters are both well-separated and internally cohesive, making it particularly useful for applications in image segmentation, social network analysis, and bioinformatics.

The Normalized Cuts algorithm is a sophisticated and effective method for spectral clustering, balancing the goals of intra-cluster cohesion and inter-cluster separation. By leveraging the eigenvectors of the normalized Laplacian, NCuts produces meaningful and balanced partitions, making it suitable for a wide range of applications.

There are several key advantages to using the Normalized Cuts algorithm:

- **Balanced clusters:** NCuts ensures that clusters are balanced in size, preventing the formation of trivial solutions where one cluster is much larger than the others.
- **Flexibility:** The algorithm can handle arbitrary shapes and sizes of clusters, unlike traditional methods that assume spherical clusters.
- **Robustness:** NCuts is robust to noise and outliers, as it considers the overall structure of the data graph rather than relying solely on local information.

Despite its strengths, the Normalized Cuts algorithm has some limitations:

- **Computational complexity:** The algorithm can be computationally intensive for very large datasets due to the eigenvector computation.
- **Parameter sensitivity:** The performance of NCuts depends on the choice of the similarity measure and its parameters, such as the bandwidth in the Gaussian kernel.
- **Scalability:** While sparse spectral clustering methods can mitigate some computational issues, scaling NCuts to extremely large datasets remains challenging.

Future research directions in spectral clustering, and specifically in the Normalized Cuts algorithm, include:

- **Scalability improvements:** Developing more efficient algorithms for eigenvector computation and sparsification techniques to handle larger datasets.
- **Parameter estimation:** Creating adaptive methods for selecting optimal parameters for the similarity measure to enhance clustering performance.
- **Application to dynamic data:** Extending NCuts to handle dynamic or streaming data, where the similarity graph evolves over time.
- **Integration with deep learning:** Combining spectral clustering with deep learning techniques to leverage feature learning and improve clustering results in high-dimensional and complex data.

In summary, the Normalized Cuts algorithm is a powerful tool in the arsenal of spectral clustering techniques. While it comes with computational challenges, its ability to produce balanced and meaningful clusters makes it a valuable method for various applications. As research continues, we can expect further enhancements in efficiency, scalability, and adaptability, broadening the scope of problems that NCuts can effectively address.

11.8   The Ratio Cuts Algorithm

One of the fundamental spectral clustering algorithms is the Ratio Cuts (RCuts) algorithm. Introduced as a means to partition a graph into balanced clusters, the Ratio Cuts algorithm aims to minimize the cut size relative to the number of nodes in each partition, ensuring that the resulting clusters are both well-separated and balanced in size. The core idea of RCuts is to represent the data as a graph where nodes correspond to data points, and edges represent the similarity between pairs of points. The objective is to partition the graph into k clusters such that the ratio of the cut (the sum of the edge weights between clusters) to the size of the clusters is minimized. This method contrasts with other partitioning strategies that might focus solely on minimizing the cut size without considering the balance between cluster sizes.

Understanding the differences between Ratio Cuts (RCuts) and Normalized Cuts (NCuts) is crucial for selecting the appropriate algorithm for a given clustering task. While both methods aim to partition a graph into meaningful clusters, they differ in their formulation of the cut objective and the balance of cluster sizes.

The primary difference between Ratio Cuts and Normalized Cuts lies in their objective functions, which define how the quality of a partition is measured.

The Ratio Cuts algorithm seeks to minimize the cut size relative to the number of nodes in each partition. The objective function for RCuts is given by:

$$\text{RCut}(A, B) = \frac{\text{cut}(A, B)}{|A|} + \frac{\text{cut}(A, B)}{|B|}$$

where $\text{cut}(A, B)$ is the sum of the weights of edges that connect nodes in partition A with nodes in partition B. Moreover, $|A|$ and $|B|$ are the sizes (number of nodes) of partitions A and B, respectively.

This formulation ensures that the algorithm not only minimizes the number of edges between clusters but also takes into account the size of each cluster, promoting balanced partitions.

The Normalized Cuts algorithm, on the other hand, aims to minimize the cut cost relative to the total edge weights associated with each partition. The objective function for NCuts is given by:

$$\text{NCut}(A, B) = \frac{\text{cut}(A, B)}{\text{assoc}(A, V)} + \frac{\text{cut}(A, B)}{\text{assoc}(B, V)}$$

where $\mathrm{assoc}(A, V)$ is the sum of the weights of all edges that connect nodes in partition A to any node in the graph V. Moreover, $\mathrm{assoc}(B, V)$ is similarly defined for partition B.

This formulation normalizes the cut value by the total degree of each partition, which helps in achieving balanced partitions while ensuring that the cut value is proportional to the internal connectivity of the clusters.

Both RCuts and NCuts algorithms strive to create balanced clusters, but they achieve this balance in different ways.

RCuts algorithm directly incorporates the size of each cluster into the objective function, explicitly penalizing partitions that result in significantly unequal cluster sizes. This makes RCuts particularly effective in scenarios where the balance of cluster sizes is critical.

NCuts algorithm achieves balance by normalizing the cut value by the total degree of each partition. This ensures that clusters are not only balanced in terms of size but also in terms of internal connectivity. As a result, the NCuts algorithm is often better suited for data with varying densities and where the internal cohesion of clusters is important.

Both RCuts and NCuts algorithms involve eigenvector computations, but their implementations can differ slightly.

The implementation of RCuts involves solving the eigenvector problem for the unnormalized Laplacian matrix. This can be computationally intensive, especially for large datasets, due to the need to compute and handle large matrices.

NCuts, in contrast, involves the normalized Laplacian matrix, which typically requires additional computational steps such as the construction of the normalized matrix. However, the normalization step can also lead to better performance in terms of producing balanced and meaningful clusters.

In summary, the main differences between Ratio Cuts and Normalized Cuts can be highlighted as follows:

- **Objective function:** RCuts minimizes the cut size relative to the number of nodes, while NCuts minimizes the cut cost relative to the total edge weights associated with each partition.
- **Balance of clusters:** RCuts explicitly balances cluster sizes, whereas NCuts balances both the size and the internal connectivity of clusters.
- **Implementation:** RCuts uses the unnormalized Laplacian, while NCuts uses the normalized Laplacian, leading to different computational requirements.

Both Ratio Cuts and Normalized Cuts are powerful spectral clustering methods, each with its strengths and ideal use cases. Understanding these differences allows practitioners to select the most suitable algorithm for their specific data clustering needs.

The Ratio Cuts algorithm provides an effective approach to spectral clustering by focusing on the balance between cluster sizes and the separation between clusters. By minimizing the cut size relative to the cluster sizes, RCuts ensures that the resulting clusters are both meaningful and well-separated.

The Ratio Cuts algorithm offers several advantages:

- **Balanced clusters:** RCuts ensures that clusters are balanced in size, avoiding trivial solutions where one cluster dominates.
- **Flexibility:** The algorithm can handle clusters of arbitrary shapes and sizes, unlike traditional methods that assume spherical clusters.
- **Robustness:** RCuts is robust to noise and outliers, as it considers the overall structure of the data graph rather than relying solely on local information.

Despite its strengths, the Ratio Cuts algorithm has some limitations:

- **Computational complexity:** The algorithm can be computationally intensive for very large datasets due to the eigenvector computation.
- **Parameter sensitivity:** The performance of RCuts depends on the choice of the similarity measure and its parameters, such as the bandwidth in the Gaussian kernel.
- **Scalability:** While sparse spectral clustering methods can mitigate some computational issues, scaling RCuts to extremely large datasets remains challenging.

Future research directions in spectral clustering, and specifically in the Ratio Cuts algorithm, include:

- **Scalability improvements:** Developing more efficient algorithms for eigenvector computation and sparsification techniques to handle larger datasets.
- **Parameter estimation:** Creating adaptive methods for selecting optimal parameters for the similarity measure to enhance clustering performance.
- **Application to dynamic data:** Extending RCuts to handle dynamic or streaming data, where the similarity graph evolves over time.

- **Integration with deep learning:** Combining spectral clustering with deep learning techniques to leverage feature learning and improve clustering results in high-dimensional and complex data.

In summary, the Ratio Cuts algorithm is a fundamental spectral clustering technique that balances cluster sizes and separations effectively. While it faces challenges in computational complexity and parameter sensitivity, ongoing research promises to address these issues and expand its applicability to broader and more complex datasets.

Bibliography

Aggarwal, C. C., Wolf, J. L., Yu, P. S., Procopiuc, C., and Park, J. S. (1999). Fast algorithms for projected clustering, in *ACM SIGMoD Record*, Vol. 28 (ACM), pp. 61–72.

Aggarwal, C. C. and Yu, P. S. (2002). Redefining clustering for high-dimensional applications, *Knowledge and Data Engineering, IEEE Transactions on* **14**, 2, pp. 210–225.

Agrawal, K., Garg, S., Sharma, S., and Patel, P. (2016). Development and validation of optics based spatio-temporal clustering technique, *Information Sciences*.

Akaike, H. (1998). Information theory and an extension of the maximum likelihood principle, in *Selected Papers of Hirotugu Akaike* (Springer), pp. 199–213.

Anand, S., Mittal, S., Tuzel, O., and Meer, P. (2014). Semi-supervised kernel mean shift clustering, *IEEE Transactions on Pattern Analysis and Machine Intelligence* **36**, 6, pp. 1201–1215.

Andrade, G., Ramos, G., Madeira, D., Sachetto, R., Ferreira, R., and Rocha, L. (2013). G-dbscan: A gpu accelerated algorithm for density-based clustering, *Procedia Computer Science* **18**, pp. 369–378.

Ankerst, M., Breunig, M. M., Kriegel, H.-P., and Sander, J. (1999). Optics: ordering points to identify the clustering structure, in *ACM Sigmod Record*, Vol. 28 (ACM), pp. 49–60.

Ardizzone, E., La Cascia, M., and Vella, F. (2008). Mean shift clustering for personal photo album organization, in *2008 15th IEEE International Conference on Image Processing* (IEEE), pp. 85–88.

Arthur, D. and Vassilvitskii, S. (2007). k-means++: The advantages of careful seeding, in *Proceedings of the eighteenth annual ACM-SIAM Symposium on Discrete algorithms* (Society for Industrial and Applied Mathematics), pp. 1027–1035.

Baker, F. B. (1974). Stability of two hierarchical grouping techniques case i: Sensitivity to data errors, *Journal of the American Statistical Association* **69**, 346, pp. 440–445.

Batagelj, V. and Zaveršnik, M. (2011). Fast algorithms for determining (generalized) core groups in social networks, *Advances in Data Analysis and Classification* **5**, 2, pp. 129–145.

Birant, D. and Kut, A. (2007). St-dbscan: An algorithm for clustering spatial-temporal data, *Data & Knowledge Engineering* **60**, 1, pp. 208–221.

Bo, S., Ding, L., Li, H., Di, F., and Zhu, C. (2009). Mean shift-based clustering analysis of multispectral remote sensing imagery, *International Journal of Remote Sensing* **30**, 4, pp. 817–827.

Bodenhofer, U., Kothmeier, A., and Hochreiter, S. (2011). Apcluster: an r package for affinity propagation clustering, *Bioinformatics* **27**, 17, pp. 2463–2464.

Bonner, R. E. (1964). On some clustering techniques, *IBM Journal of Research and Development* **8**, 1, pp. 22–32.

Borah, B. and Bhattacharyya, D. (2004). An improved sampling-based dbscan for large spatial databases, in *Intelligent Sensing and Information Processing, 2004. Proceedings of International Conference on* (IEEE), pp. 92–96.

Borgelt, C. and Kruse, R. (2004). Shape and size regularization in expectation maximization and fuzzy clustering, in *European Conference on Principles of Data Mining and Knowledge Discovery* (Springer), pp. 52–62.

Bradley, P. S., Fayyad, U., and Reina, C. (1998). Scaling em (expectation-maximization) clustering to large databases, Tech. rep., Technical Report MSR-TR-98-35, Microsoft Research Redmond.

Breunig, M. M., Kriegel, H.-P., Ng, R. T., and Sander, J. (1999). Optics-of: Identifying local outliers, in *European Conference on Principles of Data Mining and Knowledge Discovery* (Springer), pp. 262–270.

Breunig, M. M., Kriegel, H.-P., and Sander, J. (2000). Fast hierarchical clustering based on compressed data and optics, in *European Conference on Principles of Data Mining and Knowledge Discovery* (Springer), pp. 232–242.

Butts, C. T. (2010). sna: Tools for social network analysis. r package version 2.2-0.

Campello, R. J. G. B., Moulavi, D., and Sander, J. (2013). *Density-Based Clustering Based on Hierarchical Density Estimates* (Springer Berlin Heidelberg, Berlin, Heidelberg), ISBN 978-3-642-37456-2, pp. 160–172, doi:10.1007/978-3-642-37456-2_14, http://dx.doi.org/10.1007/978-3-642-37456-2_14.

Carreira-Perpiñán, M. Á. (2006). Fast nonparametric clustering with gaussian blurring mean-shift, in *Proceedings of the 23rd International Conference on Machine Learning* (ACM), pp. 153–160.

Chang, F., Qiu, W., Zamar, R. H., Lazarus, R., Wang, X., et al. (2010). Clues: an r package for nonparametric clustering based on local shrinking, *Journal of Statistical Software* **33**, 4, pp. 1–16.

Chen, M., Gao, X., and Li, H. (2010). Parallel dbscan with priority r-tree, in *Information Management and Engineering (ICIME), 2010 The 2nd IEEE International Conference on* (IEEE), pp. 508–511.

Cheng, Y. (1995). Mean shift, mode seeking, and clustering, *IEEE Transactions on Pattern Analysis and Machine Intelligence* **17**, 8, pp. 790–799.

Chipman, H. and Tibshirani, R. (2006). Hybrid hierarchical clustering with applications to microarray data, *Biostatistics* **7**, 2, pp. 286–301.

Chu, H.-K. and Lee, T.-Y. (2009). Multiresolution mean shift clustering algorithm for shape interpolation, *IEEE Transactions on Visualization and Computer Graphics* **15**, 5, pp. 853–866.

Comaniciu, D. and Meer, P. (2002). Mean shift: A robust approach toward feature space analysis, *IEEE Transactions on Pattern Analysis and Machine Intelligence* **24**, 5, pp. 603–619.

Corter, J. E. and Gluck, M. A. (1992). Explaining basic categories: Feature predictability and information. *Psychological Bulletin* **111**, 2, p. 291.

Dai, B.-R. and Lin, I.-C. (2012). Efficient map/reduce-based dbscan algorithm with optimized data partition, in *Cloud Computing (CLOUD), 2012 IEEE 5th International Conference on* (IEEE), pp. 59–66.

Darong, H. and Peng, W. (2012). Grid-based dbscan algorithm with referential parameters, *Physics Procedia* **24**, pp. 1166–1170.

Dempster, A. P., Laird, N. M., and Rubin, D. B. (1977). Maximum likelihood from incomplete data via the em algorithm, *Journal of the Royal Statistical Society. Series B (methodological)*, pp. 1–38.

Deng, Z., Hu, Y., Zhu, M., Huang, X., and Du, B. (2015). A scalable and fast optics for clustering trajectory big data, *Cluster Computing* **18**, 2, pp. 549–562.

Dhillon, I. S. and Modha, D. S. (2001). Concept decompositions for large sparse text data using clustering, *Machine Learning* **42**, 1-2, pp. 143–175.

Do-Jong, K., Yong-Woon, P., and Dong-Jo, P. (2001). A novel validity index for determination of the optimal number of clusters, *IEICE Transactions on Information and Systems* **84**, 2, pp. 281–285.

Duda, R. O., Hart, P. E., and Stork, D. G. (2012). *Pattern Classification* (John Wiley & Sons).

El Assaad, H., Samé, A., Govaert, G., and Aknin, P. (2016). A variational expectation–maximization algorithm for temporal data clustering, *Computational Statistics & Data Analysis* **103**, pp. 206–228.

Ester, M., Kriegel, H.-P., Sander, J., Xu, X., et al. (1996). A density-based algorithm for discovering clusters in large spatial databases with noise, in *Kdd*, Vol. 96, pp. 226–231.

Estivill-Castro, V. (2002). Why so many clustering algorithms: a position paper, *ACM SIGKDD Explorations Newsletter* **4**, 1, pp. 65–75.

Estivill-Castro, V. and Yang, J. (2000). Fast and robust general purpose clustering algorithms, in *PRICAI 2000 Topics in Artificial Intelligence* (Springer), pp. 208–218.

Fayyad, U., Bradley, P. S., and Reina, C. (2001). Scalable system for expectation maximization clustering of large databases, US Patent 6,263,337.

Fazayeli, F., Wang, L., and Mandziuk, J. (2008). Feature selection based on the rough set theory and expectation-maximization clustering algorithm, in *International Conference on Rough Sets and Current Trends in Computing* (Springer), pp. 272–282.

Fisher, D. H. (1987). Knowledge acquisition via incremental conceptual clustering, *Machine Learning* **2**, 2, pp. 139–172, doi:10.1007/BF00114265, http://dx.doi.org/10.1007/BF00114265.

Fortier, J. and Solomon, H. (1966). *Clustering procedures*, 62 (Academic Press New York).

Fowlkes, E. B. and Mallows, C. L. (1983). A method for comparing two hierarchical clusterings, *Journal of the American Statistical Association* **78**, 383, pp. 553–569.

Fraley, C. and Raftery, A. E. (2002). Model-based clustering, discriminant analysis, and density estimation, *Journal of the American Statistical Association* **97**, 458, pp. 611–631.

Fraley, C. and Raftery, A. E. (2007). Bayesian regularization for normal mixture estimation and model-based clustering, *Journal of Classification* **24**, 2, pp. 155–181.

Fraley, C. *et al.* (2012). mclust version 4 for r: Normal mixture modeling for model-based clustering, classification, and density estimation, *University of Washington: Seattle*.

Frawley, W. J., Piatetsky-Shapiro, G., and Matheus, C. J. (1992). Knowledge discovery in databases: An overview, *AI magazine* **13**, 3, p. 57.

Frey, B. J. and Dueck, D. (2007). Clustering by passing messages between data points, *Science* **315**, 5814, pp. 972–976.

Fu, J.-S., Liu, Y., and Chao, H.-C. (2015). Ica: An incremental clustering algorithm based on optics, *Wireless Personal Communications* **84**, 3, pp. 2151–2170.

Gengerelli, J. (1963). A method for detecting subgroups in a population and specifying their membership, *The Journal of psychology* **55**, 2, pp. 457–468.

Georgescu, B., Shimshoni, I., and Meer, P. (2003). Mean shift based clustering in high dimensions: A texture classification example, in *Computer Vision, 2003. Proceedings. Ninth IEEE International Conference on* (IEEE), pp. 456–463.

Gluck, M. (1985). Information, uncertainty and the utility of categories, in *Proc. of the Seventh Annual Conf. on Cognitive Science Society* (Lawrence Erlbaum), pp. 283–287.

Goyal, P., Kumari, S., Kumar, D., Balasubramaniam, S., and Goyal, N. (2014). Parallelizing optics for multicore systems, in *Proceedings of the 7th ACM India Computing Conference* (ACM), p. 17.

Guha, S., Rastogi, R., and Shim, K. (1998a). Cure: an efficient clustering algorithm for large databases, in *ACM SIGMOD Record*, Vol. 27 (ACM), pp. 73–84.

Guha, S., Rastogi, R., and Shim, K. (1998b). Cure: An efficient clustering algorithm for large databases, *ACM Sigmod record* **27**, 2, pp. 73–84.

Guha, S., Rastogi, R., and Shim, K. (1999). Rock: A robust clustering algorithm for categorical attributes, in *Data Engineering, 1999. Proceedings, 15th International Conference on* (IEEE), pp. 512–521.

Guo, J., Tian, D., McKinney, B. A., and Hartman IV, J. L. (2010). Recursive expectation-maximization clustering: a method for identifying buffering mechanisms composed of phenomic modules, *Chaos: An Interdisciplinary Journal of Nonlinear Science* **20**, 2, p. 026103.

Gupta, U. D., Menon, V., and Babbar, U. (2010). Detecting the number of clusters during expectation-maximization clustering using information criterion, in *Machine Learning and Computing (ICMLC), 2010 Second International Conference on* (IEEE), pp. 169–173.

Hahsler, M. (2015). dbscan: Density based clustering of applications with noise (dbscan) and related algorithms, *R package version 0.9-2*, URL *http://CRAN. R-project. org/package= dbscan*.

Han, J., Kamber, M., and Pei, J. (2011). *Data Mining: Concepts and Techniques: Concepts and Techniques* (Elsevier).

Hartigan, J. A. (1975). *Clustering Algorithms (Probability & Mathematical Statistics)* (John Wiley & Sons Inc).

He, Y., Tan, H., Luo, W., Mao, H., Ma, D., Feng, S., and Fan, J. (2011). Mr-dbscan: an efficient parallel density-based clustering algorithm using mapreduce, in *Parallel and Distributed Systems (ICPADS), 2011 IEEE 17th International Conference on* (IEEE), pp. 473–480.

Hidot, S. and Saint-Jean, C. (2010). An expectation–maximization algorithm for the wishart mixture model: Application to movement clustering, *Pattern Recognition Letters* **31**, 14, pp. 2318–2324.

Huang, Z. (1998). Extensions to the k-means algorithm for clustering large data sets with categorical values, *Data mining and knowledge discovery* **2**, 3, pp. 283–304.

Hubert, L. and Arabie, P. (1985). Comparing partitions, *Journal of Classification* **2**, 1, pp. 193–218.

Jain, A. K., Dubes, R. C., *et al.* (1988). *Algorithms for Clustering Data*, Vol. 6 (Prentice hall Englewood Cliffs).

Jain, A. K., Murty, M. N., and Flynn, P. J. (1999). Data clustering: A review, *ACM computing surveys (CSUR)* **31**, 3, pp. 264–323.

Jiang, H., Li, J., Yi, S., Wang, X., and Hu, X. (2011). A new hybrid method based on partitioning-based dbscan and ant clustering, *Expert Systems with Applications* **38**, 8, pp. 9373–9381.

Jin, X. and Han, J. (2011). Expectation maximization clustering, in *Encyclopedia of Machine Learning* (Springer), pp. 382–383.

Jollois, F.-X. and Nadif, M. (2007). Speed-up for the expectation-maximization algorithm for clustering categorical data, *Journal of Global Optimization* **37**, 4, pp. 513–525.

Jung, Y. G., Kang, M. S., and Heo, J. (2014). Clustering performance comparison using k-means and expectation maximization algorithms, *Biotechnology & Biotechnological Equipment* **28**, sup1, pp. S44–S48.

Kalita, H. K., Bhattacharya, D. K., and Kar, A. (2007). A new algorithm for ordering of points to identify clustering structure based on perimeter of triangle: Optics (bopt), in *Advanced Computing and Communications, 2007. ADCOM 2007. International Conference on* (IEEE), pp. 523–528.

Karypis, G., Han, E.-H., and Kumar, V. (1999). Chameleon: Hierarchical clustering using dynamic modeling, *Computer* **32**, 8, pp. 68–75.

Kaufman, L. and Rousseeuw, P. (1987). *Clustering by Means of Medoids* (North-Holland).

Kaufman, L. and Rousseeuw, P. J. (2009). *Finding Groups in Data: An Introduction to Cluster Analysis*, Vol. 344 (John Wiley & Sons).

King, B. (1967). Step-wise clustering procedures, *Journal of the American Statistical Association* **62**, 317, pp. 86–101.

Kishor, D. R. and Venkateswarlu, N. (2016). Hybridization of expectation-maximization and k-means algorithms for better clustering performance, *arXiv preprint arXiv:1603.07879*.

Kisilevich, S., Mansmann, F., and Keim, D. (2010). P-dbscan: a density based clustering algorithm for exploration and analysis of attractive areas using collections of geo-tagged photos, in *Proceedings of the 1st International Conference and Exhibition on Computing for Geospatial Research & Application* (ACM), p. 38.

Klosgen, W. and Zytkow, J. (2002). Kdd: The purpose, necessity and chalanges, *Handbook of Data Mining and Knowledge Discovery*, pp. 1–9.

Kriegel, H.-P., Kröoger, P., and Gotlibovich, I. (2003). Incremental optics: Efficient computation of updates in a hierarchical cluster ordering, in *International Conference on Data Warehousing and Knowledge Discovery* (Springer), pp. 224–233.

Kryszkiewicz, M. and Lasek, P. (2010). Ti-dbscan: Clustering with dbscan by means of the triangle inequality, in *International Conference on Rough Sets and Current Trends in Computing* (Springer), pp. 60–69.

Larsen, B. and Aone, C. (1999). Fast and effective text mining using linear-time document clustering, in *Proceedings of the fifth ACM SIGKDD International Conference on Knowledge Discovery and Data Mining* (ACM), pp. 16–22.

Leisch, F. (2010). Neighborhood graphs, stripes and shadow plots for cluster visualization, *Statistics and Computing* **20**, 4, pp. 457–469.

Li, X. and Bian, S. (2009). A kernel fuzzy clustering algorithm with spatial constraint based on improved expectation maximization for image segmentation, in *2009 International Conference on Measuring Technology and Mechatronics Automation*, Vol. 2 (IEEE), pp. 529–533.

Lloyd, S. P. (1982). Least squares quantization in pcm, *Information Theory, IEEE Transactions on* **28**, 2, pp. 129–137.

Luo, Q. and Khoshgoftaar, T. M. (2004). Efficient image segmentation by mean shift clustering and mdl-guided region merging, in *Tools with Artificial Intelligence, 2004. ICTAI 2004. 16th IEEE International Conference on* (IEEE), pp. 337–343.

Marcotorchino, J.-F. and Michaud, P. (1979). *Optimisation en analyse ordinale des données* (Masson).

Martinetz, T. M., Berkovich, S. G., and Schulten, K. J. (1993). Neural-gas' network for vector quantization and its application to time-series prediction, *Neural Networks, IEEE Transactions on* **4**, 4, pp. 558–569.

McLachlan, G. and Krishnan, T. (2007). *The EM Algorithm and Extensions*, Vol. 382 (John Wiley & Sons).

Murtagh, F. (1983). A survey of recent advances in hierarchical clustering algorithms, *The Computer Journal* **26**, 4, pp. 354–359.

Mustapha, N., Jalali, M., and Jalali, M. (2009). Expectation maximization clustering algorithm for user modeling in web usage mining system, *European Journal of Scientific Research* **32**, 4, pp. 467–476.

Myrvold, W. (1992). Counting k-component forests of a graph, *Networks* **22**, 7, pp. 647–652.

Nasser, S., Alkhaldi, R., and Vert, G. (2006). A modified fuzzy k-means clustering using expectation maximization, in *2006 IEEE International Conference on Fuzzy Systems* (IEEE), pp. 231–235.

Ozertem, U., Erdogmus, D., and Jenssen, R. (2008). Mean shift spectral clustering, *Pattern Recognition* **41**, 6, pp. 1924–1938.

Park, H.-S. and Jun, C.-H. (2009). A simple and fast algorithm for k-medoids clustering, *Expert Systems with Applications* **36**, 2, pp. 3336–3341.

Patwary, M. M. A., Palsetia, D., Agrawal, A., Liao, W.-k., Manne, F., and Choudhary, A. (2012). A new scalable parallel dbscan algorithm using the disjoint-set data structure, in *High Performance Computing, Networking, Storage and Analysis (SC), 2012 International Conference for* (IEEE), pp. 1–11.

Pelleg, D., Moore, A. W., et al. (2000). X-means: Extending k-means with efficient estimation of the number of clusters. in *ICML*, pp. 727–734.

Rand, W. M. (1971). Objective criteria for the evaluation of clustering methods, *Journal of the American Statistical Association* **66**, 336, pp. 846–850.

Ray, S. and Turi, R. H. (1999). Determination of number of clusters in k-means clustering and application in colour image segmentation, in *Proceedings of the 4th International Conference on Advances in Pattern Recognition and Digital Techniques* (India), pp. 137–143.

Ripley, B. D. (1996). *Pattern recognition and neural networks* (Cambridge University Press).

Rodriguez, A. and Laio, A. (2014). Clustering by fast search and find of density peaks, *Science* **344**, 6191, pp. 1492–1496.

Rong, Q.-s., Yan, J.-b., and Guo, G.-q. (2004). Research and implementation of clustering algorithm based on dbscan, *Computer Applications* **4**.

Roy, S. and Bhattacharyya, D. K. (2005). *An Approach to Find Embedded Clusters Using Density Based Techniques* (Springer Berlin Heidelberg, Berlin, Heidelberg), ISBN 978-3-540-32429-4, pp. 523–535, doi:10.1007/11604655_59, http://dx.doi.org/10.1007/11604655_59.

Ruiz, C., Spiliopoulou, M., and Menasalvas, E. (2007). C-dbscan: Density-based clustering with constraints, in *International Workshop on Rough Sets, Fuzzy Sets, Data Mining, and Granular-Soft Computing* (Springer), pp. 216–223.

Safarinejadian, B., Menhaj, M., and Karrari, M. (2009). Distributed data clustering using expectation maximization algorithm, *Journal of Applied Sciences* **9**, 5, pp. 854–864.

Schonlau, M. et al. (2002). The clustergram: A graph for visualizing hierarchical and non-hierarchical cluster analyses, *The Stata Journal* **3**, pp. 316–327.

Schwarz, G. et al. (1978). Estimating the dimension of a model, *The Annals of Statistics* **6**, 2, pp. 461–464.

Seidman, S. B. (1983). Network structure and minimum degree, *Social Networks* **5**, 3, pp. 269–287.

Selim, S. Z. and Ismail, M. A. (1984). K-means-type algorithms: a generalized convergence theorem and characterization of local optimality, *Pattern Analysis and Machine Intelligence, IEEE Transactions on*, 1, pp. 81–87.

Sharet, N. and Shimshoni, I. (2016). Analyzing data changes using mean shift clustering, *International Journal of Pattern Recognition and Artificial Intelligence* **30**, 07, p. 1650016.

Shimodaira, H. *et al.* (2004). Approximately unbiased tests of regions using multistep-multiscale bootstrap resampling, *The Annals of Statistics* **32**, 6, pp. 2616–2641.

Smiti, A. and Elouedi, Z. (2012). Dbscan-gm: An improved clustering method based on gaussian means and dbscan techniques, in *2012 IEEE 16th International Conference on Intelligent Engineering Systems (INES)* (IEEE), pp. 573–578.

Sneath, P. and Sokal, R. (1972). *Numerical Taxonomy: The Principles and Practice of Numerical Classification* (San Francisco: Freeman xvi, 573p. General (KR, 197300075)).

Špitalský, V. and Grendár, M. (2013). Optics-based clustering of emails represented by quantitative profiles, in *Distributed Computing and Artificial Intelligence* (Springer), pp. 53–60.

Strehl, A. and Ghosh, J. (2000). Clustering guidance and quality evaluation using relationship-based visualization, *Proceedings of Intelligent Engineering Systems Through Artificial Neural Networks*, pp. 483–488.

Strehl, A., Ghosh, J., and Mooney, R. (2000). Impact of similarity measures on web-page clustering, in *Workshop on Artificial Intelligence for Web Search (AAAI 2000)*, pp. 58–64.

Subbarao, R. and Meer, P. (2006). Nonlinear mean shift for clustering over analytic manifolds, in *Proceedings of the 2006 IEEE Computer Society Conference on Computer Vision and Pattern Recognition-Volume 1* (IEEE Computer Society), pp. 1168–1175.

Sutor, S., Röhr, R., Pujolle, G., and Reda, R. (2008). Efficient mean shift clustering using exponential integral kernels, *World Academic of Science, Engineering and Technology* **26**, pp. 376–380.

Tibshirani, R., Walther, G., and Hastie, T. (2001). Estimating the number of clusters in a data set via the gap statistic, *Journal of the Royal Statistical Society: Series B (Statistical Methodology)* **63**, 2, pp. 411–423.

Toth, D., Stuke, I., Wagner, A., and Aach, T. (2004). Detection of moving shadows using mean shift clustering and a significance test, in *Pattern Recognition, 2004. ICPR 2004. Proceedings of the 17th International Conference on*, Vol. 4 (IEEE), pp. 260–263.

Tran, T. N., Drab, K., and Daszykowski, M. (2013). Revised dbscan algorithm to cluster data with dense adjacent clusters, *Chemometrics and Intelligent Laboratory Systems* **120**, pp. 92–96.

Tryon, R. C. and Bailey, D. E. (1970). Cluster analysis, *New York: McGraw HillTryonCluster Analysis 1970*.

Tuzel, O., Porikli, F., and Meer, P. (2009). Kernel methods for weakly supervised mean shift clustering, in *2009 IEEE 12th International Conference on Computer Vision* (IEEE), pp. 48–55.

Veyssieres, M. and Plant, R. E. (1998). Identification of vegetation state and transition domains in california's hardwood rangelands, *University of California*, p. 101.

Viswanath, P. and Babu, V. S. (2009). Rough-dbscan: A fast hybrid density based clustering method for large data sets, *Pattern Recognition Letters* **30**, 16, pp. 1477–1488.

Viswanath, P. and Pinkesh, R. (2006). l-dbscan: A fast hybrid density based clustering method, in *18th International Conference on Pattern Recognition (ICPR'06)*, Vol. 1 (IEEE), pp. 912–915.

Wang, L., Tang, D., Guo, Y., and Do, M. N. (2015). Common visual pattern discovery via nonlinear mean shift clustering, *IEEE Transactions on Image Processing* **24**, 12, pp. 5442–5454.

Wang, X., Qiu, W., and Zamar, R. H. (2007). Clues: A non-parametric clustering method based on local shrinking, *Computational Statistics & Data Analysis* **52**, 1, pp. 286–298.

Wang, Y., Yang, J., and Peng, N. (2006). Unsupervised color–texture segmentation based on soft criterion with adaptive mean-shift clustering, *Pattern Recognition Letters* **27**, 5, pp. 386–392.

Ward Jr, J. H. (1963). Hierarchical grouping to optimize an objective function, *Journal of the American Statistical Association* **58**, 301, pp. 236–244.

Wu, K.-L. and Yang, M.-S. (2007). Mean shift-based clustering, *Pattern Recognition* **40**, 11, pp. 3035–3052.

Xiao, C. and Liu, M. (2010). Efficient mean-shift clustering using gaussian kd-tree, in *Computer Graphics Forum*, Vol. 29 (Wiley Online Library), pp. 2065–2073.

Xiaoyun, C., Yufang, M., Yan, Z., and Ping, W. (2008). Gmdbscan: multi-density dbscan cluster based on grid, in *e-Business Engineering, 2008. ICEBE'08. IEEE International Conference on* (IEEE), pp. 780–783.

Yu, Z., Jin, Y., Parmar, M., and Wang, L. (2016). Application of modified optics algorithm in e-commerce sites classification and evaluation, *Journal of Electronic Commerce in Organizations (JECO)* **14**, 1, pp. 64–75.

Yuan, X., Hu, B.-G., He, R., *et al.* (2009). Agglomerative mean-shift clustering via query set compression, in *SDM* (SIAM), pp. 223–234.

Yuan, X.-T., Hu, B.-G., and He, R. (2012). Agglomerative mean-shift clustering, *IEEE Transactions on Knowledge and Data Engineering* **24**, 2, pp. 209–219.

Yue, S.-h., Li, P., Guo, J.-d., and Zhou, S.-g. (2004). Using greedy algorithm: Dbscan revisited ii, *Journal of Zhejiang University SCIENCE* **5**, 11, pp. 1405–1412.

Zhang, X., Cui, Y., Li, D., Liu, X., and Zhang, F. (2012). An adaptive mean shift clustering algorithm based on locality-sensitive hashing, *Optik-International Journal for Light and Electron Optics* **123**, 20, pp. 1891–1894.

Zhou, A., Zhou, S., Cao, J., Fan, Y., and Hu, Y. (2000a). Approaches for scaling dbscan algorithm to large spatial databases, *Journal of Computer Science and Technology* **15**, 6, pp. 509–526.

Zhou, H., Wang, P., and Li, H. (2012). Research on adaptive parameters determination in dbscan algorithm, *Journal of Information & Computational Science* **9**, 7, pp. 1967–1973.

Zhou, S., Zhou, A., Cao, J., Wen, J., Fan, Y., and Hu, Y. (2000b). Combining sampling technique with dbscan algorithm for clustering large spatial databases, in *Pacific-Asia Conference on Knowledge Discovery and Data Mining* (Springer), pp. 169–172.

Zhou, S., Zhou, A., Jin, W., Fan, Y., and Qian, W. (2000c). Fdbscan: a fast dbscan algorithm, *Ruan Jian Xue Bao* **11**, 6, pp. 735–744.

Zhou, Y.-m., Jiang, S.-y., and Yin, M.-l. (2008). A region-based image segmentation method with mean-shift clustering algorithm, in *Fuzzy Systems and Knowledge Discovery, 2008. FSKD'08. Fifth International Conference on*, Vol. 2 (IEEE), pp. 366–370.

Index

281